CAMBRIDGE TRACTS IN MATHEMATICS
AND MATHEMATICAL PHYSICS

GENERAL EDITORS
H. BASS, J. F. C. KINGMAN, F. SMITHIES,
J. A. TODD & C. T. C. WALL

62. *Injective modules*

T0275650

D. W. SHARPE

AND

P. VÁMOS

Lecturers in Pure Mathematics
University of Sheffield

Injective modules

CAMBRIDGE

AT THE UNIVERSITY PRESS 1972

CAMBRIDGE UNIVERSITY PRESS
Cambridge, New York, Melbourne, Madrid, Cape Town, Singapore, São Paulo, Delhi

Cambridge University Press
The Edinburgh Building, Cambridge CB2 8RU, UK

Published in the United States of America by Cambridge University Press, New York

www.cambridge.org
Information on this title: www.cambridge.org/9780521083911

© Cambridge University Press 1972

First published 1972
This digitally printed version 2008

A catalogue record for this publication is available from the British Library

Library of Congress Catalogue Card Number: 79–178283

ISBN 978-0-521-08391-1 hardback
ISBN 978-0-521-09092-6 paperback

Contents

To our parents

Preface

In this book, we make no claim to give an exhaustive treatment
of the theory of injective modules. Our aim has been to see how
injective modules may be used in the context of commutative
algebra and how, by means of injective modules, the results of
commutative algebra may be generalized to apply in a non-
commutative setting. The possibility of the use of injective
modules in this way was opened up by the fundamental work of
E. Matlis in his paper 'Injective modules over Noetherian rings'.
Some of the results in this book can be obtained by more
elementary means, without the use of injective modules. But,
in our view, a good working knowledge of injective modules is
a sound investment for module theorists.

It is extremely tempting nowadays to do everything in a
general Abelian category instead of in the category of modules,
and indeed most of what has been done here fits well into that
more general setting. We have resisted this temptation. It is
still not possible to assume that every reader of a book such as
this will be familiar with the theory of Abelian categories, and
to have developed such a theory first would have seriously
unbalanced the book. We have tried to use categorical methods
where we could. Readers who are familiar with Abelian categories
will be able to adapt the results given here to the more general
setting. We refer those interested to P. Gabriel's thesis.

In Chapter 1 we merely recap on the basic notions of modules.
In Chapter 2 we introduce injective modules and construct the
injective envelope of a module. Most of the results of this chapter
will be common to any book which does more than just mention
injective modules. This chapter ends with a description of
the indecomposable injective modules over a commutative
Noetherian ring in terms of the prime ideals of the ring. This
is in preparation for Chapter 4.

Chapter 3 is concerned with semi-simplicity. The various

results in Chapters 4 and 5 on the duality that exists between Noetherian and Artinian modules have their origins in this chapter. Chapter 4 brings us to the heart of the material – the Noetherian theory. Here we deal with primary decomposition and with a characterization of Artinian modules over Noetherian rings. In Chapter 5 we show how the ring of endomorphisms of an indecomposable injective module may be used to treat localization of rings and completion of local rings. We also develop a theory of duality for complete local rings. In Chapter 6 we use the theory of injective modules to obtain complete sets of invariants for direct sum decompositions of modules.

We have added some exercises at the end of each chapter. Some of these are routine and others extend the results of the text. Also, we have included brief notes on each chapter. We have not given credit for the various results as they arise in the text; instead, we have tried to do this in the notes. We are sorry if we have failed to give credit where credit is due. We have also included a short bibliography at the end of the book. This is not intended to be complete, but rather to set down some of our sources and to suggest further reading for those who remain dissatisfied by what is offered here.

There are other books which deal with injective modules, notably the lecture notes of C. Faith and the book by J. Lambek. These two are basically concerned with the applications of injective modules to ring theory, i.e. to the construction of rings of quotients and related topics. These we have hardly touched upon. Thus there is only the obvious intersection between this book and their books.

This book arose out of a seminar given in the University of Sheffield during the session 1966/67, and we are happy to express our gratitude to the members of our audience for their patience and for their many helpful suggestions. Originally, we had intended to give only a few lectures, but Professor D. G. Northcott urged us to continue. He kindly agreed to read our manuscript and made innumerable suggestions which went well beyond routine improvements. Chapter 5 in particular was completely rewritten at his suggestion. He has been a constant encouragement to us when we have grown dispirited, and we gladly acknowledge the debt that we owe to him. At the same time, we accept full responsibility for any deficiencies that might remain.

We are also indebted to the Cambridge University Press for their patience towards us and for their unfailing helpfulness; and in particular we express our thanks to Professor C. T. C. Wall, one of the editors of this series.

D. W. SHARPE
P. VÁMOS

June 1971

Some remarks to the intending reader

One difficulty in writing a tract of this nature, where space is confined and the scope is limited, is knowing where to begin. Our subject is injective modules, but the emphasis is set squarely on the word 'injective', so much so that we must assume the reader to be familiar already with the notion of a module. This rather implies that he will have encountered rings. *Unless we state explicitly, we shall not assume that our rings are commutative, but we shall always assume that every ring has an identity element.* The symbol R will be used consistently to denote a ring without further comment. The identity element of R will be denoted by 1_R, or less explicitly by 1, and its zero element by 0_R, or just 0. We do *not* assume in general that $1_R \neq 0_R$. If $1_R = 0_R$, then the ring possesses the single element 0_R and is called a *trivial ring*; if $1_R \neq 0_R$, then R is said to be *non-trivial*.

We assume that the reader has encountered *ideals* of rings; there are three sorts, left ideals, right ideals and two-sided ideals. When I is a two-sided ideal of R, the *residue class ring* R/I can be formed; we assume that the reader knows how this can be done.

By a *module*, we shall always mean a *unitary left module*. Thus, if M is an R-module, then M is an additive Abelian group which is such that, given $r \in R$ and $m \in M$, there is defined an element rm in M; further

$$(r_1 + r_2)\, m = r_1 m + r_2 m, \quad r(m_1 + m_2) = rm_1 + rm_2,$$
$$r_1(r_2 m) = (r_1 r_2)\, m, \qquad 1_R m = m$$

for all $r, r_1, r_2 \in R$ and all $m, m_1, m_2 \in M$. We shall assume that the reader is familiar with the concepts of *homomorphism* and *isomorphism* for modules, and with *submodules* and *factor modules*.

It is at this point that, rather tentatively, we take up the story, selecting only those aspects of module theory with

which we are especially concerned. We are really expecting that much of the material in Chapter 1 will be familiar to the reader, and invite him to sample it to see if this is indeed the case. Those who need rather more background material than we have provided are referred, for example, to D. G. Northcott *Lessons on Rings, Modules and Multiplicities* (Cambridge, 1968)†
Chapter 1, where the basic notions are introduced in a leisurely fashion.

We issue a further warning. There will be a number of occasions when Zorn's Lemma is used, and the reader should be prepared for these. A short account of this is to be found in Section 2.1 of *DGN Lessons*.

There is one matter of terminology which may cause confusion unless the reader is prepared for it. A mapping $f: A \to B$, where A and B are non-empty sets, is said to be *injective* if $f(a_1) \neq f(a_2)$ whenever a_1 and a_2 are distinct elements of A. This should not be confused with the concept of an injective module.

† In the sequel, we shall refer to this book as *DGN Lessons*.

1. Preliminaries

1.1 A set-theoretical result

One of our primary tasks will be to embed a given R-module in an injective module. This will be done in Chapter 2. For many purposes, we actually want to be able to say that every module is a submodule of an injective module. This can be deduced from the result on embedding by means of a device which is essentially of a set-theoretic rather than an algebraic nature. We shall deal with this point in our first result.

PROPOSITION 1.1 *Let* $f: E \to E'$ *be an embedding*† *of the* R-*module* E *in the* R-*module* E'. *Then there is an extension module* E'' *of* E *and an isomorphism* $g: E'' \to E'$ *which extends the embedding* f, *i.e. which is such that* $g(e) = f(e)$ *for every* $e \in E$.

Proof. We first look at the set $E' \backslash f(E)$ of all elements of E' which do not belong to $f(E)$. It may happen that this set has elements in common with E, so we consider instead any set X which is an identical copy of $E' \backslash f(E)$ but is disjoint from E. There will be a bijection $\nu: X \to E' \backslash f(E)$. Put $E'' = E \cup X$. The mapping $g: E'' \to E'$ is now defined by

$$g(e'') = \begin{cases} f(e'') & \text{if} \quad e'' \in E \\ \nu(e'') & \text{if} \quad e'' \in X. \end{cases}$$

Then g is a bijection and extends f.

We now use g to give E'' the structure of an R-module in such a way that E is a submodule of E'' and g is an R-homomorphism. Let $e_1, e_2 \in E''$ and $r \in R$. Then $g(e_1), g(e_2) \in E'$, so that $g(e_1) + g(e_2)$ and $rg(e_1)$ are defined in E'. We define $e_1 + e_2$ and re_1 in E'' by

$$e_1 + e_2 = g^{-1}(g(e_1) + g(e_2)), \quad re_1 = g^{-1}(rg(e_1)). \quad (1.1.1)$$

These definitions agree with the addition and multiplication by

---

† The word 'embedding' is just another name for a monomorphism, i.e. a homomorphism which is an injective mapping.

ring elements on E, and they give E'' the structure of an R-module. It will thus be an extension module of E. Further, from (1.1.1),

$$g(e_1 + e_2) = g(e_1) + g(e_2), \quad g(re_1) = rg(e_1),$$

so that g is an R-homomorphism. $\square$

1.2 Sums and intersections of submodules

Suppose that we have an R-module M and a family $\{M_i\}_{i \in I}$ of submodules of M. The *sum*, $\sum\limits_{i \in I} M_i$, of this family is the set of all elements $\sum\limits_{i \in I} m_i$, where $m_i \in M_i$ for each $i \in I$ and $m_i = 0$ for all but a finite number of i. To include the case when I is the empty set, we define $\sum\limits_{i \in I} m_i$ when I is empty to be 0. Then $\sum\limits_{i \in I} M_i$ is a submodule of M; it is the smallest submodule of M to contain every M_i. When I is the empty set, $\sum\limits_{i \in I} M_i$ is the zero submodule of M. The largest submodule of M which is contained in every M_i is the intersection $\bigcap\limits_{i \in I} M_i$ of the family. The appropriate convention for this intersection when I is empty is that it is M. If I is a finite non-empty set, say $I = \{1, 2, \ldots, n\}$, then the sum and intersection are also written

$$M_1 + M_2 + \ldots + M_n \quad \text{and} \quad M_1 \cap M_2 \cap \ldots \cap M_n$$

respectively.

PROPOSITION 1.2 (The modular law) *Let H, K, L be submodules of an R-module M, and suppose that $K \subseteq H$. Then*

$$H \cap (K + L) = K + (H \cap L).$$

Proof. Clearly $K + (H \cap L) \subseteq H \cap (K + L)$. Consider an element h of $H \cap (K + L)$. Then $h = k + l$ for some $k \in K$, $l \in L$, and $l = h - k \in H$. Thus h belongs to $K + (H \cap L)$. This shows that $H \cap (K + L) \subseteq K + (H \cap L)$. $\square$

In technical terms, Proposition 1.2 says that the submodules of a given module form a modular lattice with respect to the operations of addition and intersection.

Now consider a subset G of the R-module M. The intersection of all submodules of M containing G will be the smallest submodule of M to contain G; it is called the submodule of M *generated* by G (or the submodule of M generated by the *elements* of G). If G is the empty set, this is just the zero submodule of M; otherwise it consists of all elements of the form

$$r_1 g_1 + r_2 g_2 + \ldots + r_n g_n,$$

where the $r_i \in R$ and the $g_i \in G$. We denote this submodule by RG.

A submodule of M which can be generated by a finite set of elements is said to be *finitely generated*; a submodule which can be generated by a single element is said to be *singly generated* or *cyclic*. For example, the ring R as an R-module is singly generated by its identity element. If $m_1, m_2, \ldots, m_n$ belong to M, then the submodule that these elements generate consists of all elements of the form

$$r_1 m_1 + r_2 m_2 + \ldots + r_n m_n,$$

where the $r_i \in R$; it is denoted by

$$Rm_1 + Rm_2 + \ldots + Rm_n.$$

If $\{M_i\}_{i \in I}$ is a family of submodules of M and if, for each i, M_i is generated by the set of elements G_i, then $\sum_{i \in I} M_i$ is generated by the union $\bigcup_{i \in I} G_i$ of the G_i. In particular, $\sum_{i \in I} M_i$ is generated by $\bigcup_{i \in I} M_i$.

A module is said to be *simple* if (i) it is non-zero and (ii) the only proper† submodule that it possesses is the zero submodule. If M is a simple R-module and if m is any non-zero element of M, then the submodule generated by m must be M itself. Thus we have the next result.

PROPOSITION 1.3 *Every simple module is singly generated.* □

Let K be a submodule of M and let A be a left ideal of R. We denote by AK the submodule of M generated by all elements of the form ak, where $a \in A$ and $k \in K$. In fact, AK is the set of all elements of the form

$$a_1 k_1 + a_2 k_2 + \ldots + a_n k_n,$$

where the $a_i \in A$ and the $k_i \in K$.

† The *proper* submodules of an R-module M are the submodules other than M itself.

1.3 Direct sums and direct products

Let $\{E_i\}_{i\in I}$ be a family of R-modules. Suppose initially that I is not empty, and consider the set E of all families $\{e_i\}$, where $e_i \in E_i$. This set can be given the structure of an R-module: we define

$$\{e_i\} + \{e_i'\} = \{e_i + e_i'\},$$

$$r\{e_i\} = \{re_i\},$$

where $r \in R$ and $e_i, e_i' \in E_i$. We call E the *direct product* of the family $\{E_i\}_{i\in I}$ and denote it by

$$\prod_{i\in I} E_i.$$

The direct product of an empty family of R-modules is defined to be a zero module.

Let E' be the subset of E consisting of all families $\{e_i\}$ for which $e_i = 0$ for all but a finite number of i. Then E' is a submodule of E. We call E' the *external direct sum* of the family $\{E_i\}_{i\in I}$ and denote it by

$$\bigoplus_{i\in I} E_i.$$

Of course, if I is a finite set then the external direct sum and the direct product coincide. If $I = \{1, 2, ..., n\}$, we may then denote the external direct sum by

$$E_1 \oplus E_2 \oplus ... \oplus E_n.$$

Let M be an R-module and let G be a generating set for M. For example, G could be the whole of M. We consider the family of R-modules indexed by G, each of the modules of the family being R itself. We denote the external direct sum of this family by $\bigoplus_{g\in G} R$. We can define a mapping

$$\phi: \bigoplus_{g\in G} R \to M$$

by

$$\phi(\{r_g\}) = \sum_{g\in G} r_g g \quad (r_g \in R).$$

Note that $r_g = 0$ for all but a finite number of g, so that ϕ is well-defined. In fact, ϕ is an R-homomorphism and also a surjective mapping, or what we shall call an *epimorphism*. We may state this as follows:

PROPOSITION 1.4 *Every R-module is a homomorphic image of an external direct sum of copies of R. If the R-module is finitely generated, then it is a homomorphic image of an external direct sum of a finite number of copies of R.* □

Let M be an R-module and let $\{M_i\}_{i \in I}$ be a family of submodules of M. We say that *the sum* $\sum\limits_{i \in I} M_i$ *is direct,* or that it is an *internal direct sum,* and we write it as

$$\sum_{i \in I} M_i \quad \text{(d.s.)},$$

if every element of $\sum\limits_{i \in I} M_i$ has a *unique* representation in the form $\sum\limits_{i \in I} m_i$, where $m_i \in M_i$ and $m_i = 0$ for all but a finite number of i. The sum of an empty family of submodules is direct.

PROPOSITION 1.5 *Let* $\{M_i\}_{i \in I}$ *be a family of submodules of an R-module M. Then the following statements are equivalent:*

(a) $M = \sum\limits_{i \in I} M_i$ *(d.s.);*

(b) $M = \sum\limits_{i \in I} M_i$ *and, for each* $j \in I$, $M_j \cap (\sum\limits_{i \neq j} M_i) = 0$.

Proof. Assume (a). Then certainly $M = \sum\limits_{i \in I} M_i$. Now consider an element j of I and an element m_j of $M_j \cap (\sum\limits_{i \neq j} M_i)$. Then we can write

$$m_j = \sum_{i \neq j} m_i,$$

where $m_i \in M_i$ and $m_i = 0$ for all but a finite number of i. Then

$$m_j + \sum_{i \neq j} (-m_i) = 0.$$

It follows from the uniqueness of sums that $m_j = 0$. This proves (b).

Now assume (b), and suppose

$$\sum_{i \in I} m_i = \sum_{i \in I} m_i',$$

where $m_i, m_i' \in M_i$ and $m_i = m_i' = 0$ for all but a finite number of i. Consider an element j of I. Then

$$m_j - m_j' = \sum_{i \neq j} (m_i' - m_i),$$

so

$$m_j - m_j' \in M_j \cap (\sum_{i \neq j} M_i) = 0.$$

Hence $m_j = m_j'$, and this is true for every j in I. This establishes (a). $\square$

Staying with the family $\{M_i\}_{i \in I}$ of submodules of M, we can form their external direct sum and can then define a mapping

$$f\colon \bigoplus_{i \in I} M_i \to \sum_{i \in I} M_i$$

by

$$f(\{m_i\}) = \sum_{i \in I} m_i,$$

where $m_i \in M_i$. This mapping is an epimorphism, and is an isomorphism if and only if the sum $\sum_{i \in I} M_i$ is direct. Thus any result about external direct sums can be transferred by means of the mapping f to a corresponding result about internal direct sums.

We return again to the family of R-modules $\{E_i\}_{i \in I}$ and denote by E' its external direct sum. For each $j \in I$, denote by E_j' the set of all elements of E' of the form $\{e_i\}$, where $e_i = 0$ if $i \neq j$. Then E_j' is a submodule of E' isomorphic to E_j and

$$E' = \sum_{i \in I} E_i' \quad \text{(d.s.)}.$$

Thus any result about internal direct sums can be transferred to a corresponding result about external direct sums.

We shall in future dispense with the adjectives 'external' and 'internal' and rely on the context to determine which type of direct sum is meant. We shall also feel entirely free to take results concerning one and apply their analogues for the other; and this we shall do without comment.

For each j in I, we can define mappings

$$\phi_j\colon E_j \to \prod_{i \in I} E_i \quad \text{and} \quad \pi_j\colon \prod_{i \in I} E_i \to E_j.$$

If $e_j \in E_j$, we put $\phi_j(e_j) = \{e_i'\}$, where $e_j' = e_j$ and $e_i' = 0$ if $i \neq j$; and $\pi_j(\{e_i\}) = e_j$, where $e_i \in E_i$. Then, for each j, ϕ_j and π_j are homomorphisms. Also, ϕ_j is an injection and π_j a surjection, so that ϕ_j is a monomorphism and π_j an epimorphism. We call the ϕ_j the *injection mappings* and the π_j the *projection mappings* of the direct product $\prod_{i \in I} E_i$. Note that, for each $j, k \in I$, the combined mapping

$$E_j \xrightarrow{\phi_j} \prod_{i \in I} E_i \xrightarrow{\pi_k} E_k$$

1.3 Direct sums and direct products 7

is the zero mapping if $j \neq k$ and is the identity mapping if $j = k$, i.e.

$$\pi_k \phi_j = 0 \text{ if } j \neq k \quad \text{and} \quad \pi_j \phi_j = \mathrm{id}_{E_j}. \tag{1.3.1}$$

Just as for direct products, so also for external and internal direct sums we can define injection and projection mappings. In the case of an internal direct sum, the injection mappings are just inclusions.

Suppose we have a finite family $\{E_i\}_{i=1}^m$ of R-modules, where $m \geqslant 1$, and denote by ϕ_i, π_i the injection and projection mappings of its direct sum E. It is easily seen that

$$\sum_{i=1}^m \phi_i \pi_i = \mathrm{id}_E. \tag{1.3.2}$$

Finally, a submodule M' of an R-module M is called a *direct summand* of M if there exists a submodule M'' of M such that

$$M = M' + M'' \quad \text{(d.s.)}.$$

1.4 Some applications of Zorn's Lemma to modules

DEFINITION   *Let M be an R-module. A submodule K of M is said to be a 'maximal submodule' of M if* (i) *K is a proper submodule of M and* (ii) *there is no proper submodule of M strictly containing K.*

PROPOSITION 1.6   *Let M be a finitely generated R-module. Then every proper submodule of M is contained in a maximal submodule of M.*

Proof. Let M' be a proper submodule of M, and denote by Ω the collection of all proper submodules of M which contain M'. Then Ω is not empty, because M' belongs to Ω. We may partially order Ω by inclusion. Let Σ be a non-empty totally ordered subset of Ω, and denote by M_0 the union of all the members of Σ. Then $M_0 \supseteq M'$, and it may be verified that M_0 is a submodule of M. Note that this verification uses the fact that Σ is totally ordered. Moreover, M_0 is a proper submodule of M. For suppose otherwise, and let $\{m_1, m_2, ..., m_n\}$ be a set of generators of M. Then each m_i belongs to a member of Σ, so there must be a member of Σ which contains every m_i, i.e. which is the whole of M. This is just not so. Thus $M_0 \in \Omega$ and is an upper bound of Σ. It follows by Zorn's Lemma that Ω possesses a maximal member K (say), i.e. M has a maximal submodule containing M'. $\square$

COROLLARY 1 *A non-zero finitely generated R-module possesses a maximal submodule.*

Proof. Apply Proposition 1.6 to the zero submodule. $\square$

COROLLARY 2 *Every proper left ideal of R is contained in a maximal left ideal.*

Proof. This follows from Proposition 1.6 since R is a singly generated module over itself, when its left ideals become submodules. $\square$

It is worth noting that the proof of Proposition 1.6 breaks down if M is not finitely generated, because we cannot then be sure that M_0 is proper. For a similar sort of reason, Zorn's Lemma will not give the existence of a 'minimal' submodule of an arbitrary non-zero module.

There will be a number of occasions when we shall need to use Zorn's Lemma in situations involving direct sums, and although the contexts will be different in each case the application of the Lemma is the same. It is convenient to deal with the situation here.

Suppose we concentrate our attention on some collection $\mathscr{A}$ of submodules of M. For example, $\mathscr{A}$ could be the collection of all simple submodules of M. We denote by Ω the collection of subsets of $\mathscr{A}$ which have the property that the sum of the submodules in the subset is direct. For example, a collection containing a single submodule from $\mathscr{A}$ would belong to Ω; but certainly the empty set belongs to Ω. Then Ω is not empty and may be partially ordered by inclusion. Let Σ be a non-empty totally ordered subset of Ω. Consider the union $\mathscr{A}'$ of all members of Σ. We wish to show that $\mathscr{A}' \in \Omega$, i.e. the sum of the submodules in $\mathscr{A}'$ is direct. Put $\mathscr{A}' = \{M_i\}_{i \in I}$, and suppose that

$$\sum_{i \in I} m_i = \sum_{i \in I} m'_i,$$

where $m_i, m'_i \in M_i$ and $m_i = m'_i = 0$ for all but a finite number of i. Suppose that the values of i for which m_i and m'_i are not both zero are $i_1, i_2, \ldots, i_n$. Then

$$m_{i_1} + m_{i_2} + \ldots + m_{i_n} = m'_{i_1} + m'_{i_2} + \ldots + m'_{i_n}.$$

Now each of $M_{i_1}, M_{i_2}, \ldots, M_{i_n}$ belongs to some member of Σ and Σ is totally ordered, so there is a member of Σ which contains all of $M_{i_1}, M_{i_2}, \ldots, M_{i_n}$. This must mean that $m_{i_\alpha} = m'_{i_\alpha}$ for $\alpha = 1, 2, \ldots, n$. It follows that $m_i = m'_i$ for every value of i, so that the sum of the submodules in $\mathscr{A}'$ is direct, i.e. $\mathscr{A}' \in \Omega$. Thus Σ is bounded above, so, by Zorn's Lemma, Ω has a maximal member.

For reference, we state this conclusion in a proposition.

PROPOSITION 1.7 *Let $\mathscr{A}$ be a collection of submodules of an R-module M. Then there is a maximal collection of members of $\mathscr{A}$ whose sum is direct.* $\square$

1.5 The isomorphism theorems

Let $f: M \to N$ be a homomorphism of R-modules, let A be a submodule of M and let B be a submodule of N. Then $f(A)$ is a submodule of N contained in $\operatorname{Im} f$, the image of f, and $f^{-1}(B)$, which consists of all elements m of M such that $f(m) \in B$, is a submodule of A containing $\operatorname{Ker} f$, the kernel of f. Also,

$$f^{-1}(f(A)) = A + \operatorname{Ker} f \quad \text{and} \quad f(f^{-1}(B)) = B \cap \operatorname{Im} f,$$

so that, if $A \supseteq \operatorname{Ker} f$ then $f^{-1}(f(A)) = A$, and if $B \subseteq \operatorname{Im} f$ then $f(f^{-1}(B)) = B$. This gives the next result.

PROPOSITION 1.8 *With the above notation, there is a one–one correspondence between the submodules of M containing $\operatorname{Ker} f$ and the submodules of N contained in $\operatorname{Im} f$. This is such that, if $A (\supseteq \operatorname{Ker} f)$ and $B (\subseteq \operatorname{Im} f)$ correspond, then $B = f(A)$ and $A = f^{-1}(B)$. This correspondence preserves inclusion.* $\square$

The last remark of Proposition 1.8 just expresses the fact that, if A_1, A_2 are submodules of M such that $A_1 \subseteq A_2$, then

$$f(A_1) \subseteq f(A_2),$$

and if B_1, B_2 are submodules of N such that $B_1 \subseteq B_2$, then $f^{-1}(B_1) \subseteq f^{-1}(B_2)$.

Note also that, if $\{A_i\}_{i \in I}$ is a non-empty family of submodules of M, each of which contains $\operatorname{Ker} f$, then

$$f(\bigcap_{i \in I} A_i) = \bigcap_{i \in I} f(A_i). \tag{1.5.1}$$

We consider again an R-module M with a submodule A. We assume that the reader is familiar with the construction of the factor module M/A. A typical element of M/A is a subset of M of the form

$$m + A = \{m + a : a \in A\},$$

where $m \in M$, and $m_1 + A = m_2 + A$ $(m_1, m_2 \in M)$ if and only if $m_1 - m_2 \in A$. The operations on M/A are given by

$$(m_1 + A) + (m_2 + A) = (m_1 + m_2) + A$$

and $$r(m + A) = (rm) + A,$$

where $r \in R$ and $m, m_1, m_2 \in M$. We can define a mapping

$$\phi : M \to M/A$$

by $\phi(m) = m + A$ $(m \in M)$. This mapping is an epimorphism and has kernel A; it is called the *natural mapping* of M on M/A. Note that, if B is a submodule of M, then $\phi(B) = (A + B)/A$. Thus, if $B \supseteq A$, $\phi(B) = B/A$.

We now recast Proposition 1.8 in terms of the natural mapping of M on M/A.

PROPOSITION 1.9   *Let A be a submodule of an R-module M. Then there is a one–one correspondence between the submodules of M containing A and the submodules of M/A. This is such that, if B is a submodule of M containing A, then B corresponds to B/A. This correspondence preserves inclusion.* $\square$

We note also that, if $\{A_i\}_{i \in I}$ is a non-empty family of submodules of M containing A, then

$$(\bigcap_{i \in I} A_i)/A = \bigcap_{i \in I} (A_i/A). \qquad (1.5.2)$$

This is just (1.5.1) applied to the natural mapping.

Let $f : M \to N$ be a homomorphism of R-modules, let A be a submodule of M and B a submodule of N. Suppose further that $f(A) \subseteq B$. Then we can define a mapping

$$f^* : M/A \to N/B$$

by $f^*(m + A) = f(m) + B$ $(m \in M)$. It may be verified that this does define a mapping and that this mapping is an R-homomorphism. We refer to f^* as the mapping *induced* by f.

PROPOSITION 1.10   *With the above notation, the induced mapping from M/A to N/B is an epimorphism if f is an epimorphism and is a monomorphism if $A = f^{-1}(B)$.*

Proof. Easy.☐

We single out three special cases of this result, called the *isomorphism theorems for modules*. The symbol $\approx$ will be used here and subsequently to denote an isomorphism between modules.

COROLLARY 1   *Let $f: M \to N$ be an epimorphism of R-modules. Then the mapping f induces an isomorphism*

$$M/\mathrm{Ker}\, f \approx N,$$

where the elements $m + \mathrm{Ker}\, f\, (m \in M)$ and $f(m)$ correspond under this isomorphism.

Proof. Put $B = 0$ and $A = \mathrm{Ker}\, f$. Note also that $N/0 \approx N$ under the obvious isomorphism.☐

COROLLARY 2   *Let A, B be submodules of an R-module M, with $A \subseteq B$. Then*

$$M/B \approx (M/A)/(B/A),$$

where the elements $m + B\, (m \in M)$ and $(m + A) + (B/A)$ correspond under this isomorphism.

Proof. Let $\phi: M \to M/A$ denote the natural mapping. Then B/A is a submodule of M/A and $\phi^{-1}(B/A) = B$.☐

COROLLARY 3   *Let A, B be submodules of an R-module M. Then*

$$B/(A \cap B) \approx (A + B)/A,$$

where the elements $b + (A \cap B)$ and $b + A$ correspond under this isomorphism, b being an arbitrary element of B.

Proof. Consider the inclusion mapping $f: B \to A + B$. Then $f^{-1}(A) = A \cap B$, so there is induced a monomorphism

$$B/(A \cap B) \to (A + B)/A.$$

Since every element of $(A + B)/A$ is of the form $b + A$, where $b \in B$, this mapping is actually an isomorphism.☐

As an illustration of the use of Proposition 1.10 Corollary 1, we have the following result:

PROPOSITION 1.11 *Suppose that the R-module M is the direct sum of submodules M_1 and M_2. Then*

$$M/M_1 \approx M_2, \quad M/M_2 \approx M_1.$$

Proof. The projections $M \to M_2$ and $M \to M_1$ have respective kernels M_1 and M_2. □

1.6 Annihilators and change of ring

Let K be a submodule of an R-module E, and let L be a non-empty subset of E. We denote by $K:L$ the set of all elements r of R such that $rl \in K$ for every $l \in L$, i.e.

$$K:L = \{r \in R : rL \subseteq K\}.$$

Then $K:L$ *is a left ideal of R. If L itself is a submodule of E, then $K:L$ is a two-sided ideal of R.* Similarly, if S is a non-empty subset of R, we put

$$K:_E S = \{e \in E : Se \subseteq K\}.$$

If S is a right ideal of R, then $K:_E S$ is a submodule of E which contains K. Note that

$$K:_E R = K.$$

There are various modifications of this notation. For example, if $e \in E$ we put

$$K:e = \{r \in R : re \in K\}.$$

If E' is a submodule of E, then $0:E'$ is called the *annihilator of E'*; it may also be denoted by $\mathrm{Ann}_R E'$. If $e \in E$, then $0:e$ is called the *annihilator of e*. We note that, *if E_1 and E_2 are isomorphic modules, then $\mathrm{Ann}_R E_1 = \mathrm{Ann}_R E_2$. Also, if $e_1 \in E_1$ and $e_2 \in E_2$ are elements which correspond under such an isomorphism, then $0:e_1 = 0:e_2$.*

If I is a *two-sided* ideal of R, then R/I is an R-module and

$$\mathrm{Ann}_R(R/I) = I.$$

It follows that, if I_1 and I_2 are *two-sided* ideals of R, then

$$R/I_1 \approx R/I_2 \text{ implies that } I_1 = I_2. \tag{1.6.1}$$

Consider a singly generated R-module Re. There is a mapping

$$f: R \to Re$$

given by $f(r) = re$ $(r \in R)$, and this is an epimorphism of R-modules. The kernel of f is just $0 : e$, so Proposition 1.10 Corollary 1 gives

$$Re \approx R/(0 : e), \qquad (1.6.2)$$

where the elements re $(r \in R)$ and $r + (0 : e)$ correspond under this isomorphism.

We can now describe the simple R-modules.

PROPOSITION 1.12   *Every R-module of the form R/M, where M is a maximal left ideal of R, is simple. Further, every simple R-module is isomorphic to R/M for some maximal left ideal M of R.*

Proof. The first remark follows from Proposition 1.9. Now suppose that S is a simple R-module. By Proposition 1.3, we can write $S = Re$ for some $e \in S$, so by (1.6.2) $S \approx R/(0 : e)$. Again by Proposition 1.9, $0 : e$ must be a maximal left ideal of R.☐

Let I be a *two-sided* ideal of R. Then R/I has the structure not merely of an R-module but also of a ring, the *residue class ring of R modulo I*. Since

$$r_1(r_2 + I) = (r_1 r_2) + I = (r_1 + I)(r_2 + I) \quad (r_1, r_2 \in R),$$

we see that *the submodules of R/I when R/I is considered as an R-module and the left ideals of R/I are one and the same*.

We continue to suppose that I is a two-sided ideal of R. If E is an R-module such that $\mathrm{Ann}_R E \supseteq I$, then E can be given the structure of an (R/I)-module: we put

$$(r + I)e = re \quad (r \in R, e \in E). \qquad (1.6.3)$$

(In particular, E can be given the structure of an $(R/\mathrm{Ann}_R E)$-module.) Conversely, if E' is an (R/I)-module, then E' can be given the structure of an R-module: we put

$$re' = (r + I)e' \quad (r \in R, e' \in E'). \qquad (1.6.4)$$

Moreover, when this is done, $0 :_R E' \supseteq I$. This gives the basic result concerning change of ring:

PROPOSITION 1.13   *Let I be a two-sided ideal of R. Then the R-modules whose annihilators contain I and the (R/I)-modules are one and the same. Further, if E is such a module, then its submodules are the same whether it is considered as an R-module or as an*

(R/I)-*module. If* E_1, E_2 *are two such modules, then a mapping* $E_1 \to E_2$ *is an R-homomorphism if and only if it is an* (R/I)-*homomorphism.*

Proof. The last two assertions follow straight from the definitions (1.6.3) and (1.6.4).□

There is one further remark concerning annihilators that will be used in Chapter 4. Once it has been grasped, it becomes trivial.

PROPOSITION 1.14 *Let E be an R-module and let* I_1, I_2 *be two-sided ideals of R such that* $I_1 \supseteq I_2$. *Then* $0:_E I_2$ *may be considered as an* (R/I_2)-*module, and*

$$0:_{(0:_E I_2)}(I_1/I_2) = 0:_E I_1. \;\square$$

1.7 Zero sequences and exact sequences

Let A, B, C be R-modules, and suppose we have homomorphisms as shown:

$$A \xrightarrow{\phi} B \xrightarrow{\psi} C. \tag{1.7.1}$$

If $\operatorname{Im}\phi \subseteq \operatorname{Ker}\psi$, so that $\psi\phi(a) = 0$ for all $a \in A$, then (1.7.1) is said to be a *zero sequence*. If $\operatorname{Im}\phi = \operatorname{Ker}\psi$, then (1.7.1) is said to be an *exact sequence*. Thus, to say that the sequence

$$0 \to A \xrightarrow{\phi} B \tag{1.7.2}$$

is exact is to say that $\operatorname{Ker}\phi = 0$, i.e. ϕ is a monomorphism; to say that the sequence

$$B \xrightarrow{\psi} C \to 0 \tag{1.7.3}$$

is exact is to say that $\operatorname{Im}\psi = C$, i.e. ψ is an epimorphism. Combining (1.7.1), (1.7.2) and (1.7.3), we say that the sequence

$$0 \to A \xrightarrow{\phi} B \xrightarrow{\psi} C \to 0$$

is exact if (i) ϕ is a monomorphism, (ii) ψ is an epimorphism and (iii) $\operatorname{Im}\phi = \operatorname{Ker}\psi$. For example, if E' is a submodule of an R-module E, then

$$0 \to E' \to E \to E/E' \to 0$$

is an exact sequence, where $E' \to E$ is the inclusion mapping and $E \to E/E'$ is the natural mapping. As a further example, consider a direct sum $E_1 \oplus E_2$ of R-modules. Then the sequence

$$0 \to E_1 \to (E_1 \oplus E_2) \to E_2 \to 0$$

is exact, where $E_1 \to (E_1 \oplus E_2)$ is the injection mapping and $(E_1 \oplus E_2) \to E_2$ is the projection mapping.

1.8 Noetherian and Artinian modules

It will frequently be necessary to impose certain finiteness conditions on our rings and modules.

DEFINITION *An R-module E is said to satisfy the 'maximal (resp. minimal) condition for submodules' if every non-empty collection Ω of submodules possesses a maximal (resp. minimal) member, i.e. Ω has a member K_0 such that there is no member of Ω strictly containing (resp. contained in) K_0.*

DEFINITION *An R-module E is said to satisfy the 'ascending chain condition' (resp. 'descending chain condition') if every ascending (resp. descending) chain of submodules*

$$K_1 \subseteq K_2 \subseteq K_3 \subseteq \dots \ (resp. \ K_1 \supseteq K_2 \supseteq K_3 \supseteq \dots)$$

terminates (i.e. there exists a positive integer m such that $K_n = K_m$ whenever $n \geqslant m$).

PROPOSITION 1.15 *Let E be an R-module. Then the following statements are equivalent:*
(a) E satisfies the maximal (resp. minimal) condition for submodules;
(b) E satisfies the ascending (resp. descending) chain condition.
Proof. We shall prove that the maximal condition for submodules is equivalent to the ascending chain condition. Minor modifications of the proof will give that the minimal condition for submodules is equivalent to the descending chain condition.

Suppose first that E satisfies the maximal condition for submodules, and consider the ascending chain

$$K_1 \subseteq K_2 \subseteq K_3 \subseteq \dots$$

of submodules of E. Then there is among the K_is a maximal element, say K_m. Clearly $K_n = K_m$ whenever $n \geqslant m$, so that the chain terminates.

Now suppose that there is a non-empty collection Ω of submodules of E without a maximal member. Consider any member K_1 of Ω. Since K_1 is not maximal, there exists K_2 in Ω such that $K_1 \subset K_2$. (The symbol $\subset$ is reserved for strict inclusion, so that $K_1 \subseteq K_2$ but $K_1 \neq K_2$.) Since K_2 is not maximal, there exists K_3 in Ω such that $K_2 \subset K_3$. If we continue in this way, we generate an infinite strictly ascending chain of submodules of E and E does not satisfy the ascending chain condition. □

DEFINITION   *A module which satisfies the maximal condition for submodules or, equivalently, the ascending chain condition is said to be 'Noetherian'. A ring is said to be 'left Noetherian' if it is Noetherian as a left module over itself.*

DEFINITION   *A module which satisfies the minimal condition for submodules or, equivalently, the descending chain condition is said to be 'Artinian'. A ring is said to be 'left Artinian' if it is Artinian as a left module over itself.*

The definitions of right Noetherian and Artinian rings are analogous to those of left Noetherian and Artinian rings. Since the theory is being developed here in terms of left modules, we shall refer simply to Noetherian and Artinian rings, meaning left Noetherian and left Artinian rings respectively.

A simple module is both Noetherian and Artinian, as is a zero module. Also, the property of being a Noetherian (resp. Artinian) module is preserved under isomorphism.

PROPOSITION 1.16   *Let E be an R-module. Then the following statements are equivalent:*

(a) *E is Noetherian;*

(b) *every submodule of E is finitely generated.*

Proof. Suppose first that E is Noetherian, and consider a submodule K of E. Denote by Ω the collection of all finitely generated submodules of K. Then Ω is not empty, because it contains the zero submodule. Since E is Noetherian, Ω will have a maximal member K_0 (say). Put $K_0 = Rk_1 + \ldots + Rk_n$ and suppose that $K_0 \neq K$. Then there exists $k \in K$, $k \notin K_0$. But then

$$K_0 + Rk = Rk_1 + \ldots + Rk_n + Rk$$

is a member of Ω strictly containing K_0. It follows that $K_0 = K$, so that K is finitely generated.

Now suppose that every submodule of E is finitely generated, and consider an ascending chain

$$K_1 \subseteq K_2 \subseteq K_3 \subseteq \ldots$$

of submodules of E. Put $K = \bigcup_{i=1}^{\infty} K_i$. Then K is a submodule of E and so is finitely generated, say $K = Rk_1 + \ldots + Rk_r$. Now each k_j belongs to one of the K_is, so there exists m such that $k_1, \ldots, k_r$ belong to K_m. But then $K = K_m$, and $K_n = K_m$ whenever $n \geqslant m$. This shows that E is Noetherian. $\square$

REMARK   As things stand here, the duality between Noetherian and Artinian modules is rather deficient, in that we cannot provide an analogue to Proposition 1.16 for Artinian modules. This deficiency will be remedied in Theorem 3.21.

PROPOSITION 1.17   *Let*

$$0 \to E' \to E \to E'' \to 0$$

be an exact sequence of R-modules and R-homomorphisms. Then E is Noetherian (resp. Artinian) if and only if both E' and E'' are Noetherian (resp. Artinian).

Proof. We shall consider only the Noetherian case, since the proof of the Artinian case is similar. We may suppose for the proof that E' is a submodule of E and that $E'' = E/E'$.

Suppose first that E is Noetherian. Since the submodules of E' are also submodules of E, it is immediate that E' is Noetherian. Now consider an ascending chain of submodules

$$K_1'' \subseteq K_2'' \subseteq K_3'' \subseteq \ldots$$

of E/E'. In the manner of Proposition 1.9, this corresponds to a chain
$$K_1 \subseteq K_2 \subseteq K_3 \subseteq \ldots$$

of submodules of E which contain E'. Thus there exists m such that $K_n = K_m$ whenever $n \geqslant m$. Returning to E/E', we see that $K_n'' = K_m''$ whenever $n \geqslant m$. Thus E/E' is Noetherian.

Now suppose that E' and E/E' are Noetherian, and consider an ascending chain of submodules

$$K_1 \subseteq K_2 \subseteq K_3 \subseteq \ldots \qquad (1.8.1)$$

of E. This gives ascending chains of submodules

$$K_1 \cap E' \subseteq K_2 \cap E' \subseteq K_3 \cap E' \subseteq \ldots,$$

$$(K_1 + E')/E' \subseteq (K_2 + E')/E' \subseteq (K_3 + E')/E' \subseteq \ldots$$

of E' and E/E' respectively. Both of these chains terminate, so there exists m such that

$$K_n \cap E' = K_m \cap E' \quad \text{and} \quad K_n + E' = K_m + E'$$

whenever $n \geqslant m$. (A further use of Proposition 1.9 has been made at this point.) Then, for every $n \geqslant m$,

$$\begin{aligned} K_n &= K_n \cap (K_n + E') \\ &= K_n \cap (K_m + E') \\ &= K_m + (K_n \cap E') \text{ by Proposition 1.2} \\ &= K_m + (K_m \cap E') \\ &= K_m. \end{aligned}$$

Thus the ascending chain (1.8.1) terminates and E is Noetherian. $\square$

PROPOSITION 1.18 *Let $E_1, E_2, \ldots, E_r$ be R-modules. Then $E_1 \oplus E_2 \oplus \ldots \oplus E_r$ is Noetherian (resp. Artinian) if and only if each E_i is Noetherian (resp. Artinian).*

Proof. We consider only the Noetherian case, since the proof of the Artinian case is similar. We use induction on r, the result being trivial when $r = 1$ (or even when $r = 0$). Now let $r > 1$, and suppose the result known for $r - 1$. We have an exact sequence

$$0 \to (E_1 \oplus \ldots \oplus E_{r-1}) \to (E_1 \oplus \ldots \oplus E_r) \to E_r \to 0, \quad (1.8.2)$$

where the mappings are the injection and projection mappings. Suppose that $E_1 \oplus \ldots \oplus E_r$ is Noetherian. By Proposition 1.17,

$E_1 \oplus \ldots \oplus E_{r-1}$ and E_r are Noetherian, and the inductive hypothesis gives that each E_i is Noetherian. Conversely, suppose that each E_i is Noetherian. By the inductive hypothesis,

$$E_1 \oplus \ldots \oplus E_{r-1}$$

is Noetherian, and Proposition 1.17 applied to the exact sequence (1.8.2) shows that $E_1 \oplus \ldots \oplus E_r$ is Noetherian. $\square$

COROLLARY *Let $S_1, S_2, \ldots, S_r$ be simple R-modules. Then $S_1 \oplus S_2 \oplus \ldots \oplus S_r$ is both Noetherian and Artinian.* $\square$

PROPOSITION 1.19 *The following statements are equivalent:*
(a) *R is a Noetherian (resp. Artinian) ring;*
(b) *every finitely generated R-module is Noetherian (resp. Artinian).*

Proof. Since R is singly generated as a module over itself, (b) implies (a). On the other hand, let E be a finitely generated R-module. By Proposition 1.4, E is a homomorphic image of a direct sum of a finite number of copies of R, and (b) follows from (a) by Propositions 1.18 and 1.17. $\square$

PROPOSITION 1.20 *A non-zero Artinian module possesses a simple submodule.*

Proof. Let E be a non-zero Artinian module. Then the collection of all non-zero submodules of E has a minimal member, and this will be a simple submodule of E. $\square$

PROPOSITION 1.21 *Let R be an Artinian ring. Then every non-zero R-module has a simple submodule.*

Proof. Let E be a non-zero R-module and consider any non-zero element e of E. By Proposition 1.19, Re is Artinian and so possesses a simple submodule (Proposition 1.20). This will, of course, be a simple submodule of E as well. $\square$

Let R be a ring and I a two-sided ideal of R. Then we can form the residue class ring R/I. Since the R-submodules of R/I are the same as its left ideals, *to say that R/I is a Noetherian (resp. Artinian) R-module is the same as to say that R/I is a Noetherian (resp. Artinian) ring.* Proposition 1.17 now gives the following result:

PROPOSITION 1.22 *If R is a Noetherian (resp. Artinian) ring, then so is the residue class ring R/I.* $\square$

1.9   The Hom functor

Let E and F be R-modules. We denote by

$$\mathrm{Hom}_R(E, F)$$

the set of all R-homomorphisms of E into F. The set $\mathrm{Hom}_R(E, F)$ may be given the structure of an *Abelian group*. If

$$f_1, f_2 \in \mathrm{Hom}_R(E, F),$$

we define $f_1 + f_2$ by

$$(f_1 + f_2)(e) = f_1(e) + f_2(e) \quad (e \in E).$$

It is an easy matter to verify that $f_1 + f_2 \in \mathrm{Hom}_R(E, F)$ and that this binary operation turns $\mathrm{Hom}_R(E, F)$ into an Abelian group; its zero element is the zero mapping $E \to F$. If either $E = 0$ or $F = 0$, then $\mathrm{Hom}_R(E, F)$ is a trivial group, i.e. $\mathrm{Hom}_R(E, F) = 0$.

Suppose that R is a *commutative* ring. Then $\mathrm{Hom}_R(E, F)$ may be given the structure of an R-module: if $r \in R$ and $f \in \mathrm{Hom}_R(E, F)$, we define rf by

$$(rf)(e) = rf(e) \quad (e \in E).$$

Then $rf \in \mathrm{Hom}_R(E, F)$. The verification of this last remark uses the fact that R is commutative.

We return to a general ring R and an R-module E. We call the R-homomorphisms from E into itself, i.e. the elements of $\mathrm{Hom}_R(E, E)$, *endomorphisms* of E. The Abelian group

$$\mathrm{Hom}_R(E, E)$$

can be given the structure of a ring: if $f_1, f_2 \in \mathrm{Hom}_R(E, E)$, we define $f_1 f_2 \in \mathrm{Hom}_R(E, E)$ by

$$f_1 f_2(e) = f_1(f_2(e)) \quad (e \in E).$$

In other words, the multiplication in $\mathrm{Hom}_R(E, E)$ is just composition of mappings. This operation is associative and distributive over addition, and the identity mapping of E acts as identity element of the ring. We note that this ring is non-trivial if and only if $E \neq 0$. We call this *the ring of endomorphisms of E*.

PROPOSITION 1.23   *The ring of endomorphisms of a simple module is a division ring.*

Proof. Let E be a simple R-module. Then $E \neq 0$, so that its ring of endomorphisms is non-trivial. Let f be a non-zero element of $\text{Hom}_R(E, E)$. Then $f(E)$ is a non-zero submodule of E and so must be E itself. Also, $\text{Ker } f$ is a submodule of E, it cannot be E, so it must be zero. Thus f is an isomorphism and so has an inverse. $\square$

For the next result, we consider two finite families $\{E_i\}_{i=1}^m$ and $\{F_j\}_{j=1}^n$ of R-modules. Put

$$E = \bigoplus_{i=1}^m E_i, \quad F = \bigoplus_{j=1}^n F_j,$$

and let $\phi_i: E_i \to E, \quad \pi_i: E \to E_i \quad (1 \leqslant i \leqslant m)$

and $\phi_j': F_j \to F, \quad \pi_j': F \to F_j \quad (1 \leqslant j \leqslant n)$

be the appropriate injections and projections. We can define mappings

$$\Lambda: \text{Hom}_R(E, F) \to \bigoplus_{\substack{1 \leqslant i \leqslant m \\ 1 \leqslant j \leqslant n}} \text{Hom}_R(E_i, F_j)$$

and $$\Lambda': \bigoplus_{\substack{1 \leqslant i \leqslant m \\ 1 \leqslant j \leqslant n}} \text{Hom}_R(E_i, F_j) \to \text{Hom}_R(E, F)$$

by $$\Lambda(f) = (\dots, \pi_j' f \phi_i, \dots) \quad (f \in \text{Hom}_R(E, F))$$

and

$$\Lambda'(\dots, f_{ij}, \dots) = \sum_{\substack{1 \leqslant k \leqslant m \\ 1 \leqslant l \leqslant n}} \phi_l' f_{kl} \pi_k \quad (f_{ij} \in \text{Hom}_R(E_i, F_j)).$$

Then, if $f \in \text{Hom}_R(E, F)$,

$$\Lambda'\Lambda(f) = \Lambda'(\dots, \pi_j' f \phi_i, \dots)$$

$$= \sum_{\substack{1 \leqslant k \leqslant m \\ 1 \leqslant l \leqslant n}} \phi_l' \pi_l' f \phi_k \pi_k$$

$$= \left(\sum_{l=1}^n \phi_l' \pi_l' \right) f \left(\sum_{k=1}^m \phi_k \pi_k \right)$$

$$= f \text{ by } (1.3.2).$$

Also, if $f_{ij} \in \text{Hom}_R(E_i, F_j)$ for $1 \leqslant i \leqslant m$, $1 \leqslant j \leqslant n$,

$$\Lambda\Lambda'(\dots, f_{ij}, \dots) = \Lambda(\sum_{\substack{1 \leqslant k \leqslant m \\ 1 \leqslant l \leqslant n}} \phi_l' f_{kl} \pi_k)$$

$$= (\dots, \pi_j'(\sum_{\substack{1 \leqslant k \leqslant m \\ 1 \leqslant l \leqslant n}} \phi_l' f_{kl} \pi_k) \phi_i, \dots)$$

$$= (\dots, f_{ij}, \dots) \text{ by } (1.3.1).$$

Thus Λ and Λ' are inverse mappings, so that they are both bijections.

It is easily checked that Λ is a homomorphism of Abelian groups, i.e. a Z-homomorphism, where Z denotes the ring of integers. Indeed, if R is a commutative ring, then Λ is actually an R-homomorphism. We have now established the following result:

PROPOSITION 1.24 *Let* $\{E_i\}_{i=1}^m$ *and* $\{F_j\}_{j=1}^n$ *be finite families of R-modules. Then there is a Z-isomorphism*

$$\mathrm{Hom}_R\left(\overset{m}{\underset{i=1}{\oplus}} E_i, \overset{n}{\underset{j=1}{\oplus}} F_j\right) \approx \underset{\substack{1 \leqslant i \leqslant m \\ 1 \leqslant j \leqslant n}}{\oplus} \mathrm{Hom}_R(E_i, F_j).$$

Moreover, if R is a commutative ring, this isomorphism is actually an R-isomorphism.□

Suppose we are given an R-homomorphism $\phi\colon F_1 \to F_2$. Then we can define a mapping

$$\Phi\colon \mathrm{Hom}_R(E, F_1) \to \mathrm{Hom}_R(E, F_2)$$

by $\Phi(f) = \phi f$, where $f \in \mathrm{Hom}_R(E, F_1)$. Now $\mathrm{Hom}_R(E, F_1)$ and $\mathrm{Hom}_R(E, F_2)$ are Abelian groups, or Z-modules, Z again denoting the ring of integers, and Φ is a Z-homomorphism. Indeed, if R is commutative, then $\mathrm{Hom}_R(E, F_1)$ and $\mathrm{Hom}_R(E, F_2)$ are actually R-modules and Φ is an R-homomorphism. Thus, from the sequence

$$0 \to F_1 \overset{\phi}{\to} F_2 \overset{\psi}{\to} F_3 \tag{1.9.1}$$

of R-modules and R-homomorphisms we can derive the sequence

$$0 \to \mathrm{Hom}_R(E, F_1) \overset{\Phi}{\to} \mathrm{Hom}_R(E, F_2) \overset{\Psi}{\to} \mathrm{Hom}_R(E, F_3) \tag{1.9.2}$$

of Abelian groups and Z-homomorphisms, where Ψ is defined in an analogous way to Φ.

PROPOSITION 1.25 *If the sequence* (1.9.1) *is exact, then so is the sequence* (1.9.2).

REMARK In functorial language, this may be described by the statement that *the functor Hom is covariant and left exact in its second variable.*

Proof. Suppose that the sequence (1.9.1) is exact. If

$$f \in \operatorname{Hom}_R(E, F_1),$$

then $\Psi\Phi(f) = \psi\phi f = 0$, so that (1.9.2) is a zero sequence. Also, if $\Phi(f) = 0$, then $\phi f = 0$, from which it follows that $f = 0$, since ϕ is a monomorphism. Thus Φ is a monomorphism. It remains to show that $\operatorname{Ker}\Psi \subseteq \operatorname{Im}\Phi$. Suppose that $g \in \operatorname{Ker}\Psi$. Then $\psi g = 0$. We wish to find $g' \in \operatorname{Hom}_R(E, F_1)$ such that $\phi g' = g$. Let $e \in E$. Then $\psi g(e) = 0$, so that $g(e) \in \operatorname{Ker}\psi = \operatorname{Im}\phi$. Thus there exists $x \in F_1$ such that $g(e) = \phi(x)$. Note that there is only one such element x, because ϕ is a monomorphism. We now define $g'(e) = x$. Then $\phi g'(e) = \phi(x) = g(e)$. Since this is true for all $e \in E$, this gives $\phi g' = g$. It only remains to show that g' is an R-homomorphism. We leave this to the reader. $\square$

Now suppose that we are given an R-homomorphism $\theta: E_1 \to E_2$. We can define a mapping

$$\Theta: \operatorname{Hom}_R(E_2, F) \to \operatorname{Hom}_R(E_1, F)$$

by $\Theta(f) = f\theta$, where $f \in \operatorname{Hom}_R(E_2, F)$. This mapping is a Z-homomorphism; indeed, if R is commutative, it is actually an R-homomorphism. Thus, from the sequence

$$E_1 \xrightarrow{\theta} E_2 \xrightarrow{\sigma} E_3 \to 0 \tag{1.9.3}$$

of R-modules and R-homomorphisms we can derive the sequence

$$0 \to \operatorname{Hom}_R(E_3, F) \xrightarrow{\Sigma} \operatorname{Hom}_R(E_2, F) \xrightarrow{\Theta} \operatorname{Hom}_R(E_1, F) \tag{1.9.4}$$

of Z-modules and Z-homomorphisms, where Σ is defined in an analogous way to Θ.

PROPOSITION 1.26 *If the sequence* (1.9.3) *is exact, then so is the sequence* (1.9.4).

REMARK In functorial language, this may be described by the statement that *Hom is contravariant and left exact in its first variable.* The word 'contravariant' is used to indicate that between (1.9.3) and (1.9.4) there has been a reversal of arrows.

Proof. Suppose that the sequence (1.9.3) is exact. If

$$f \in \operatorname{Hom}_R(E_3, F),$$

then $\Theta\Sigma(f) = f\sigma\theta = 0$, so that (1.9.4) is a zero sequence. Also, if $\Sigma(f) = 0$, then $f\sigma = 0$, from which it follows that $f = 0$, since σ is an epimorphism. Thus Σ is a monomorphism.

It remains to show that $\mathrm{Ker}\,\Theta \subseteq \mathrm{Im}\,\Sigma$. Suppose that $g \in \mathrm{Ker}\,\Theta$. Then $g\theta = 0$. We wish to find $g' \in \mathrm{Hom}_R(E_3, F)$ such that $g'\sigma = g$. Let $e_3 \in E_3$. Then there exists $e_2 \in E_2$ such that $\sigma(e_2) = e_3$. We define $g'(e_3) = g(e_2)$. It is not at all clear that we can do this, so let e_2' be another element of E_2 such that $\sigma(e_2') = e_3$. Then

$$e_2 - e_2' \in \mathrm{Ker}\,\sigma = \mathrm{Im}\,\theta,$$

so there exists $e_1 \in E_1$ such that $\theta(e_1) = e_2 - e_2'$. Then

$$g(e_2) - g(e_2') = g\theta(e_1) = 0,$$

so $g(e_2) = g(e_2')$. Hence g' is a well-defined mapping and $g'\sigma = g$. It only remains to show that g' is an R-homomorphism. We leave this to the reader. $\square$

If we refer back to the sequence (1.9.1), it is conspicuous that no claim has been made that exactness of the sequence

$$F_2 \to F_3 \to 0$$

implies exactness of the sequence

$$\mathrm{Hom}_R(E, F_2) \to \mathrm{Hom}_R(E, F_3) \to 0.$$

The class of modules E which have this property is just the class of *projective modules*. Nor in (1.9.3) have we claimed that exactness of the sequence
$$0 \to E_1 \to E_2$$

implies exactness of the sequence

$$\mathrm{Hom}_R(E_2, F) \to \mathrm{Hom}_R(E_1, F) \to 0.$$

The class of modules F which have this property is just the class of *injective modules*. It is to these modules that we now turn our attention.

Exercises on Chapter 1

1.1 Let R be a commutative domain with quotient field K. Show how K may be given the structure of an R-module. Show that, when K is finitely generated as an R-module, R must be a field.

1.2 Let R be a commutative domain, let A be a proper ideal of R and let B be a finitely generated ideal of R such that $B = AB$. Show that $B = 0$.

1.3 Let R be a ring with a unique maximal left ideal and let E be an R-module such that every finitely generated submodule of E is cyclic. Show that the submodules of E are totally ordered (i.e. given submodules A and B of E, then either $A \subseteq B$ or $B \subseteq A$).

1.4 Let $\{M_i\}_{i \in I}$ be a family of non-zero submodules of a finitely generated R-module M such that

$$M = \sum_{i \in I} M_i \ (\text{d.s.}).$$

Show that the index set I is finite.

1.5 Let $\{M_i\}_{i=1}^{\infty}$ be a family of submodules of an R-module M. Show that the sum $\sum_{i=1}^{\infty} M_i$ is direct if and only if, for every $n > 1$,

$$M_n \cap (M_1 + M_2 + \ldots + M_{n-1}) = 0.$$

1.6 Prove Lemmas 5.7, 5.8, 5.9, 5.10.

1.7 Give an example of a module which is Noetherian but not Artinian.

1.8 Let Z denote the ring of integers and Q the rational numbers considered as a Z-module. Let p be a prime number and let A be the set of all rational numbers of the form n/p^k, where $n \in Z$ and k is a non-negative integer. Show that A is a submodule of Q which contains Z and that A/Z is Artinian but not Noetherian.

1.9 Let R be a commutative ring with the property that the residue class ring R/I is Noetherian for every non-zero ideal I of R. Show that R is Noetherian. Give an example to show that this does not hold with 'Noetherian' replaced by 'Artinian'.

1.10 Let R be a commutative ring and let A be an R-module. Show that

(i) when A is Noetherian, then the residue class ring $R/\text{Ann}_R A$ is a Noetherian ring;

(ii) when A is finitely generated and Artinian, then $R/\text{Ann}_R A$ is an Artinian ring. [*Hint:* Put $A = Ra_1 + Ra_2 + \ldots + Ra_s$ and consider a particular mapping from R to $Ra_1 \oplus Ra_2 \oplus \ldots \oplus Ra_s$.]

1.11 Let R be a ring and denote by $R[X]$ the ring of all polynomials in the indeterminate X with coefficients in R. For a left ideal A of $R[X]$ and a non-negative integer n, define $I_n(A)$ to be the set of all r in R for which there is a polynomial in A of the form
$$rX^n + r_{n-1}X^{n-1} + \ldots + r_1 X + r_0.$$
Show that $I_n(A)$ is a left ideal of R.

Let A, B be left ideals of R and m, n be non-negative integers. Prove that

(i) when $A \subseteq B$, then $I_n(A) \subseteq I_n(B)$,

(ii) when $m \leqslant n$, then $I_m(A) \subseteq I_n(A)$,

(iii) when $A \subseteq B$ and $I_s(A) = I_s(B)$ for all non-negative integers s, then $A = B$.

1.12 (*The Hilbert basis theorem*) Let R be a left Noetherian ring. Show that the polynomial ring $R[X]$ is left Noetherian. [*Hint:* Use Exercise 1.11.] Give an example to show that this does not hold with 'Noetherian' replaced by 'Artinian'.

Notes on Chapter 1

Originally it was ideals in rings that were studied. Then it was seen that there were advantages to be gained from the greater generality provided by modules. There are several excellent accounts of the basics of module theory – see for example [22]. More recently, it has been recognized that the correct setting for what is done here is an Abelian category, which allows the greatest generality whilst retaining the essential properties of modules which are contained in this chapter. The interested reader is referred to [7], [9] and [21].

2. *Injective modules and injective envelopes*

Notation

We recall that the symbol R denotes a ring with an identity element. This ring need not be commutative unless it is explicitly stated. The symbol Z will be used consistently to denote the ring of integers.

We shall have frequent recourse to diagrams of R-modules and R-homomorphisms. When a diagram such as Fig. 2.1 appears, with a dotted arrow, it means that the mapping $B \to C$ is not in the first instance given and is supplied in the course of the subsequent discussion. Thus the dotted arrow should be ignored as far as the

Fig. 2.1

data is concerned. This device will avoid a multiplicity of diagrams, and merely accords with one's normal practice of filling in various pieces of information on a diagram as they are discovered.

Various other notational shorthands will appear in diagrams. Thus

$\xrightarrow{\text{inj}}$ will indicate an *injection*,

$\xrightarrow{\text{proj}}$ will indicate a *projection*,

both of these related to a direct sum or a direct product,

$\xrightarrow{\text{inc}}$ will indicate an *inclusion mapping*,

$\xrightarrow{\text{id}}$ will indicate an *identity mapping*,

$\xrightarrow{\approx}$ will indicate an *isomorphism*.

In a diagram of R-modules (resp. Z-modules) and mappings between them, it will be understood that the mappings are R-homomorphisms (resp. Z-homomorphisms). We shall not feel it necessary to point this out on each occasion.

2.1   Injective modules

PROPOSITION 2.1   *Let E be an R-module. Then the following statements are equivalent*:

(a) *given any diagram Fig. 2.2 of R-modules (and R-homomorphisms), where the row is exact, there exists an R-homomorphism $B \to E$ such that the resulting diagram (shown) is commutative*;

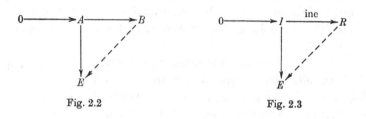

Fig. 2.2   Fig. 2.3

(b) *given any diagram Fig. 2.3 where I is a left ideal of R, there exists an R-homomorphism $R \to E$ such that the resulting diagram (shown) is commutative*;

(c) *given any exact sequence*

$$0 \to A \to B \to C \to 0$$

of R-modules, the sequence

$$0 \to \mathrm{Hom}_R(C, E) \to \mathrm{Hom}_R(B, E) \to \mathrm{Hom}_R(A, E) \to 0$$

(see Section 1.9) is exact.

DEFINITION   *An R-module E which satisfies the equivalent conditions (a), (b), (c) is said to be 'injective'. An Abelian group is said to be 'injective' if it is injective as a Z-module.*

REMARK   The property of being an injective module is preserved under isomorphism. Also, nothing is lost in (a) if it is taken that A is a submodule of B and that $A \to B$ is the inclusion mapping.

Proof of Proposition 2.1. By Proposition 1.26, (c) is equivalent to the assertion that, if the sequence

$$0 \to A \overset{\theta}{\to} B$$

is exact, then so is the sequence

$$\operatorname{Hom}_R(B, E) \overset{\Theta}{\to} \operatorname{Hom}_R(A, E) \to 0,$$

where the mapping Θ is given by $\Theta(f) = f\theta$ ($f \in \operatorname{Hom}_R(B, E)$). This amounts to saying that, if we have a diagram such as Fig. 2.4 where the row is exact, then there exists

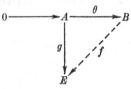

$$f \in \operatorname{Hom}_R(B, E)$$

such that $f\theta = \Theta(f) = g$. But this is exactly what is meant by the statement that Fig. 2.4 is commutative, so (c) and (a) are equivalent.

Fig. 2.4

We now assume (b) and deduce (a). Since (b) is but a special case of (a), this will complete the proof. We begin with the unaugmented diagram given in (a), where we assume that A is a submodule of B and that $A \to B$ is the inclusion mapping. We denote the given mapping $A \to E$ by h.

We denote by Ω the collection of all pairs (C, ϕ), where (i) C is a submodule of B containing A, and (ii) $\phi: C \to E$ is an R-homomorphism such that the diagram Fig. 2.5 is commutative, i.e. ϕ extends the mapping h to C. Then Ω is not empty, because

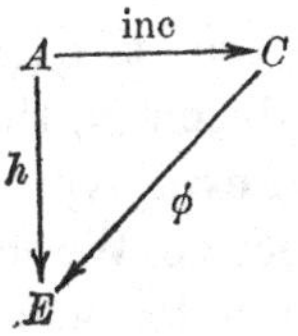

Fig. 2.5

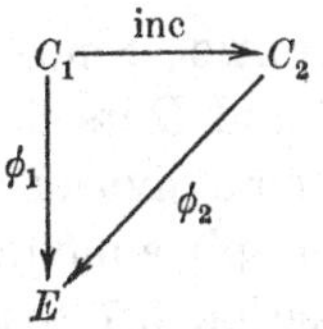

Fig. 2.6

$(A, h) \in \Omega$. We introduce a partial order on Ω. We shall say that $(C_1, \phi_1) \leqslant (C_2, \phi_2)$ if C_1 is a submodule of C_2 and Fig. 2.6 is commutative. Then Ω is an inductive system. For if $\{(C_\lambda, \phi_\lambda)\}_{\lambda \in \Lambda}$ is a totally ordered subset of Ω, then it has upper bound $(\bar{C}, \bar{\phi}) \in \Omega$, where $\bar{C} = \bigcup_{\lambda \in \Lambda} C_\lambda$ and $\bar{\phi}: \bar{C} \to E$ is defined by $\bar{\phi}(\bar{c}) = \phi_\lambda(\bar{c})$ if $\bar{c} \in C_\lambda$. Hence, by Zorn's Lemma, Ω has a maximal member (C_0, ϕ_0), say. Now (a) will follow if we show that $C_0 = B$.

Suppose that $C_0 \neq B$. Then there exists $x \in B$, $x \notin C_0$. Put $I = C_0 : x$, i.e. I consists of all elements r of R such that $rx \in C_0$. Then I is a left ideal of R. We define $\mu : I \to E$ by $\mu(r) = \phi_0(rx)$, where $r \in I$. Then μ is an R-homomorphism and, by (b), there exists an R-homomorphism $\nu : R \to E$ such that the diagram Fig. 2.7 is commutative. We can now define a mapping

$$\psi : C_0 + Rx \to E$$

by $\psi(c_0 + rx) = \phi_0(c_0) + \nu(r) \quad (c_0 \in C_0, \, r \in R).$

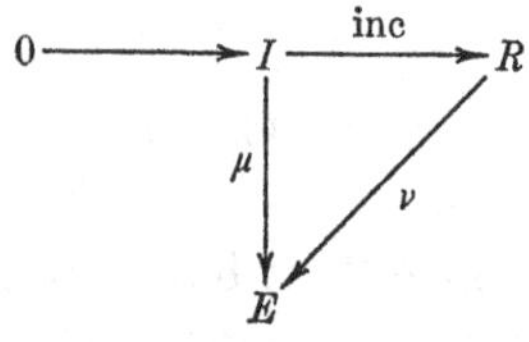

Fig. 2.7 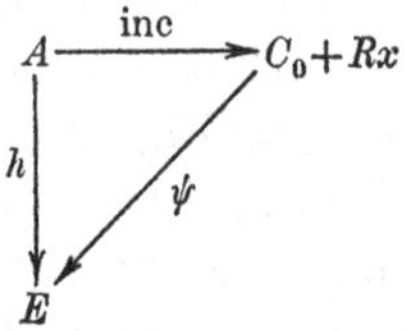   Fig. 2.8

If $c_0 + rx = 0$, then r belongs to $C_0 : x$, which is just I, and

$$\phi_0(c_0) + \nu(r) = -\phi_0(rx) + \nu(r) = -\mu(r) + \nu(r) = 0.$$

It follows that ψ is a well-defined R-homomorphism. But, if $a \in A$, then $\psi(a) = \phi_0(a) = h(a)$, so that $(C_0 + Rx, \psi) \in \Omega$. But $(C_0, \phi_0) < (C_0 + Rx, \psi)$. This contradicts the maximality of (C_0, ϕ_0) and (a) is established. $\square$

EXAMPLES   Since a division ring D has only two left ideals, namely 0 and D itself, it follows using condition (b) that every module over a division ring is injective. Thus every vector space, which is simply a module over a field, is injective. We shall show in Proposition 3.7 that the class of rings R for which every R-module is injective is precisely the class of semi-simple rings. Note that every zero module is injective.

PROPOSITION 2.2   *Let $\{E_\lambda\}_{\lambda \in \Lambda}$ be a family of R-modules. Then $\prod\limits_{\lambda \in \Lambda} E_\lambda$ is injective if and only if each E_λ is injective.*

Proof. Put $E = \prod\limits_{\lambda \in \Lambda} E_\lambda$, and denote the injections and projections associated with this direct product by

$$\phi_\lambda : E_\lambda \to E, \quad \pi_\lambda : E \to E_\lambda$$

respectively.

Suppose first that each E_λ is injective, and consider Fig. 2.9 where the row is exact. For each λ, this gives rise to Fig. 2.10. There exists $g_\lambda \in \mathrm{Hom}_R(B, E_\lambda)$ such that the resulting diagram is commutative. We now define $g \colon B \to E$ by

$$g(b) = \{g_\lambda(b)\}_{\lambda \in \Lambda} \quad (b \in B). \tag{2.1.1}$$

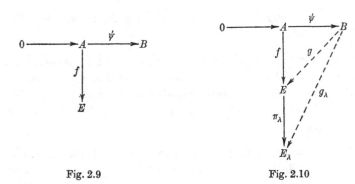

Fig. 2.9   Fig. 2.10

Then g is an R-homomorphism and, if $a \in A$,

$$g\psi(a) = \{g_\lambda \psi(a)\} = \{\pi_\lambda f(a)\} = f(a),$$

which shows that E is injective.

Conversely, suppose that E is injective, let $\lambda \in \Lambda$ and consider Fig. 2.11 where the row is exact. This gives rise to Fig. 2.12. There exists $h \in \mathrm{Hom}_R(B, E)$ such that $h\psi = \phi_\lambda \mu$. We now define $h' \colon B \to E_\lambda$ by

$$h'(b) = \pi_\lambda h(b) \quad (b \in B).$$

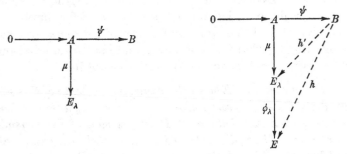

Fig. 2.11   Fig. 2.12

Then h' is an R-homomorphism and, if $a \in A$,

$$h'\psi(a) = \pi_\lambda h\psi(a) = \pi_\lambda \phi_\lambda \mu(a) = \mu(a)$$

by (1.3.1). This shows that each E_λ is injective. $\square$

PROPOSITION 2.3 *Let $\{E_\lambda\}_{\lambda \in \Lambda}$ be a family of R-modules.*
(i) *If $\underset{\lambda \in \Lambda}{\oplus} E_\lambda$ is injective, then each E_λ is injective.*
(ii) *If the index set Λ is finite and each E_λ is injective, then $\underset{\lambda \in \Lambda}{\oplus} E_\lambda$ is injective.*

Proof. Both parts of this result follow from Proposition 2.2 once we recall that the direct sum and the direct product of a family consisting of a finite number of R-modules coincide. Part (ii) is immediate. In (i), we note that each E_λ is a direct summand of $\underset{\lambda \in \Lambda}{\oplus} E_\lambda$. $\square$

We cannot in general relax the condition in (ii) that the index set should be finite. The argument given in Proposition 2.2 breaks down for finite sums at (2.1.1), because we cannot guarantee that only a finite number of the $g_\lambda(b)$ are non-zero. Indeed, this provides a characterization of left Noetherian rings. We shall show in Theorem 4.1 that the ring R is left Noetherian if and only if the direct sum of an arbitrary family of injective R-modules is injective.

2.2 Divisibility

We recall that an element r of R is said to be a *right (resp. left) zero-divisor* if there exists s in R such that $s \neq 0$ but $sr = 0$ (resp. $rs = 0$). Of course, if R is commutative, there is no difference between a right and a left zero-divisor, and we refer simply to a *zero-divisor* of R. The distinctive features of a *domain* over and above a ring are that it is non-trivial and has no zero-divisors, right or left, except, of course, the zero element itself.

DEFINITION *Let E be an R-module. An element e of E is said to be 'divisible' if, for every r of R which is not a right zero-divisor, there exists $e' \in E$ such that $e = re'$. If every element of E is divisible, then E is said to be a 'divisible module'. An Abelian group is said to be 'divisible' if it is divisible as a Z-module. Alternatively, E is* *divisible if E = rE whenever r is an element of R which is not a right zero-divisor.*

As a comparison, let E_1 be a right R-module and let e_1 be an element of E_1. Then e_1 is said to be divisible if, for every r_1 of R which is not a left zero-divisor, there exists $e_1' \in E_1$ such that $e_1 = e_1' r_1$.

EXAMPLES   The field of fractions of a commutative domain, when considered as a module over that domain in the obvious fashion, is a divisible module. In particular, the rational numbers form a divisible group under addition. Also, every module over a division ring is divisible, so every vector space is divisible. Every zero module is divisible.

LEMMA 2.4   *Let E be a divisible R-module and let E' be a submodule of E. Then E/E' is a divisible R-module.*
Proof. Consider the elements $e + E'$ $(e \in E)$ of E/E' and $r \in R$, where r is not a right zero-divisor. Then there exists $e' \in E$ such that $e = re'$, which implies that $e + E' = r(e' + E')$. Thus E/E' is divisible.☐

LEMMA 2.5   *Let $\{E_\lambda\}_{\lambda \in \Lambda}$ be a family of divisible R-modules. Then $\prod_{\lambda \in \Lambda} E_\lambda$ and $\bigoplus_{\lambda \in \Lambda} E_\lambda$ are divisible R-modules.*
Proof. Let $\{e_\lambda\} \in \prod_{\lambda \in \Lambda} E_\lambda$, where $e_\lambda \in E_\lambda$, and let r be an element of R which is not a right zero-divisor. Then, for each λ in Λ, there exists $e_\lambda' \in E_\lambda$ such that $e_\lambda = re_\lambda'$. Then $\{e_\lambda\} = r\{e_\lambda'\}$, so that $\prod_{\lambda \in \Lambda} E_\lambda$ is divisible. If $\{e_\lambda\} \in \bigoplus_{\lambda \in \Lambda} E_\lambda$, then we can arrange for $\{e_\lambda'\}$ to belong to $\bigoplus_{\lambda \in \Lambda} E_\lambda$ by insisting that $e_\lambda' = 0$ if $e_\lambda = 0$.☐

PROPOSITION 2.6   *Every injective module is divisible.*†
Proof. Let E be an injective R-module, let $e \in E$ and let r be an element of R which is not a right zero-divisor. Consider Fig. 2.13 where

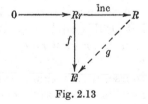

Fig. 2.13

---

† Or, as paraphrased by one of our students, 'every dejected module is invisible'!

$f: Rr \to E$ is defined by $f(sr) = se$ $(s \in R)$. Note that, because r is not a right zero-divisor, if $sr = 0$ then $s = 0$ and $f(sr) = 0$, so f is a well-defined R-homomorphism. Hence there exists a homomorphism $g: R \to E$ which agrees with f on Rr. Thus

$$e = f(r) = g(r) = rg(1),$$

which shows that E is divisible. $\square$

We shall now look for partial converses to Proposition 2.6.

DEFINITION *Let R be a domain and let E be an R-module. An element e of E is said to be a 'torsion element' of E if there exists a non-zero element r of R such that $re = 0$. If the only torsion element of E is the zero element, then E is said to be 'torsion-free'.*

PROPOSITION 2.7 *Let R be a commutative domain and let E be a torsion-free divisible R-module. Then E is injective.*

Proof. Consider Fig. 2.14 where I is an ideal of R and $f: I \to E$ is an R-homomorphism. This can obviously be completed if I is the zero ideal, so we assume that $I \neq 0$. Consider a non-zero element s of I. Because E is divisible, there exists $e \in E$ such that $f(s) = se$. Let t be any element of I. Then

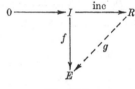

Fig. 2.14

$$sf(t) = f(st) = f(ts) = tf(s) = tse = ste,$$

whence $f(t) = te$ because E is torsion-free. We now define the R-homomorphism $g: R \to E$ by $g(r) = re$ $(r \in R)$. Then g agrees with f on I, so that E is injective. $\square$

EXAMPLE It follows from Proposition 2.7 that the field of fractions of a commutative domain is an injective module over that domain.

A domain R is said to be a *left principal ideal domain* if every left ideal is singly generated, i.e. if every left ideal of R is of the form Rr for some $r \in R$. A *right principal ideal domain* is a domain with the property that every right ideal is singly generated. Of course, if R is commutative, the distinction between left and right is unnecessary, and we shall refer simply to a *principal ideal domain*. The ring Z of integers is an example of a principal ideal domain.

THEOREM 2.8 *Let R be a left principal ideal domain and let E be an R-module. Then E is injective if and only if it is divisible.*

Proof. We assume that E is divisible and show that it is injective. By Proposition 2.6, this will complete the proof. Consider Fig. 2.15 where I is a left ideal of R. We must complete the diagram with an R-homomorphism $g: R \to E$, and for this it is sufficient to consider the case when $I \neq 0$. Now R is a left principal ideal domain, so we can write $I = Rs$ for some non-zero element s of I. Because E is divisible, there exists

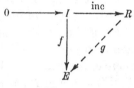

Fig. 2.15

$e \in E$ such that $f(s) = se$. We now define the R-homomorphism $g: R \to E$ by $g(r) = re$ $(r \in R)$. Then, if $t \in R$,

$$g(ts) = tse = tf(s) = f(ts),$$

so that g agrees with f on I. $\square$

COROLLARY *An Abelian group is injective if and only if it is divisible.* $\square$

Now let R be a commutative domain with field of fractions K and consider K as an R-module in the usual way. Thus the product of the elements a of R and b/c of K, where $b, c \in R$ and $c \neq 0$, is just $(ab)/c$. Let F be an R-submodule of K such that there is a non-zero element r of R such that $rF \subseteq R$. Such a submodule of K is called a *fractional ideal* of R. In this context the ideals of R are termed *integral ideals* of R. Clearly every integral ideal is a fractional ideal. If we put $I = rF$, then I is an integral ideal of R and $F = (1/r)I$.

Let F, F' be fractional ideals of R. We denote by FF' the set of all finite sums $\sum_i f_i f'_i$, where $f_i \in F$ and $f'_i \in F'$. This is also a fractional ideal of R, and the operation on the set of fractional ideals so defined is commutative and associative. Moreover, $RF = F$ for all fractional ideals F, so that R acts as a neutral element. If, given a fractional ideal F, there is a fractional ideal F' such that $FF' = R$, then F' is unique and is called the *inverse* of F; in this case F is said to be *invertible*. For example, the non-zero principal ideal Rr of R has inverse $R(1/r)$. Note that $R(1/r)$ is a fractional ideal of R since $r(R(1/r)) \subseteq R$.

DEFINITION *A 'Dedekind domain' is a commutative domain with the property that every non-zero fractional ideal is invertible.*

LEMMA 2.9 *Let R be a commutative domain such that every non-zero integral ideal is invertible. Then R is a Dedekind domain.* Proof. Let F be a non-zero fractional ideal. Then there is a non-zero element r of R such that rF is an integral ideal, whence there exists a fractional ideal F' such that $(rF)\,F' = R$. But then $F(rF') = R$ and F has inverse rF', which is a fractional ideal. $\square$

EXAMPLE Every commutative principal ideal domain is a Dedekind domain.

PROPOSITION 2.10 *Every divisible module over a Dedekind domain is injective.*

Proof. Let R be a Dedekind domain and let E be a divisible R-module. Further, let I be an integral ideal of R and consider the diagram Fig. 2.16. We must produce an R-homomorphism $g \colon R \to E$ which extends the R-homomorphism f, and for this it is sufficient to consider the case $I \neq 0$. Then I has a fractional inverse F (say), so that $IF = R$. Thus there exist elements

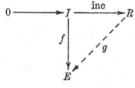

Fig. 2.16

$s_1, \dots, s_n$ in I and $k_1, \dots, k_n$ in F such that $\sum\limits_{i=1}^{n} s_i k_i = 1$. We may suppose that no s_i is zero. Now E is divisible, so for each i $(1 \leqslant i \leqslant n)$ there exists $e_i \in E$ such that $f(s_i) = s_i e_i$. Consider $s \in I$. Since $sk_i \in IF = R$,

$$f(s) = f\!\left(\sum_{i=1}^{n} sk_i s_i \right) = \sum_{i=1}^{n} (sk_i) f(s_i) = s \sum_{i=1}^{n} k_i s_i e_i.$$

Now $k_i s_i \in FI = R$, so that $\sum\limits_{i=1}^{n} k_i s_i e_i \in E$ and we can define an R-homomorphism $g \colon R \to E$ by

$$g(r) = r \sum_{i=1}^{n} k_i s_i e_i \quad (r \in R).$$

Then g agrees with f on I and we have proved that E is injective. $\square$

There is a converse to Proposition 2.10, namely that if R is a commutative domain such that every divisible R-module is injective, then R is a Dedekind domain. This is the substance of Theorem 4.25. It is in anticipation of that proof that Lemma 2.9 has been inserted. This provides a characterization of Dedekind domains.

2.3 The embedding theorem

Before injective modules can prove to be useful objects of study, we must know that there are enough of them. This is ensured by the main result of this section.

THEOREM 2.11 *Every module can be embedded in an injective module.*

The proof will be divided up into a number of lemmas.

LEMMA 2.12 *Every Abelian group can be embedded in an injective Abelian group.*

Proof. Let G be an Abelian group, and put

$$F = \bigoplus_{g \in G} Z, \quad F' = \bigoplus_{g \in G} Q,$$

where Q denotes the additive group of rational numbers. Define $\phi: F \to G$ by

$$\phi(\{n_g\}_{g \in G}) = \sum_{g \in G} n_g g,$$

where $n_g \in Z$ and $n_g = 0$ for all but a finite number of g. Then ϕ is a Z-epimorphism, so we have an isomorphism $G \approx F/K$, where $K = \operatorname{Ker} \phi$ (Proposition 1.10 Corollary 1). Hence G is isomorphic to a subgroup of F'/K. But Q is a divisible group, so by Lemmas 2.5 and 2.4 F'/K is divisible. Theorem 2.8 Corollary now gives that F'/K is injective, and we have embedded G in the injective Abelian group F'/K.□

If G is an Abelian group, then we can form the Abelian group $\operatorname{Hom}_Z (R, G)$. This can be given the structure of an R-module: if $r \in R$ and $f \in \operatorname{Hom}_Z (R, G)$, we define $rf \in \operatorname{Hom}_Z (R, G)$ by

$$(rf)(r') = f(r'r) \quad (r' \in R).$$

LEMMA 2.13 *Let G be an injective Abelian group. Then $\operatorname{Hom}_Z (R, G)$ is an injective R-module.*

Proof. We consider the diagram of R-modules Fig. 2.17, where the row is exact. A further diagram of Z-modules and Z-homomorphisms may be constructed as in Fig. 2.18. We define $\phi: A \to G$ by

$$\phi(a) = (f(a))\,(1),$$

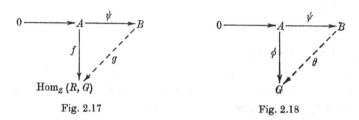

Fig. 2.17   Fig. 2.18

where $a \in A$. Hence there exists a Z-homomorphism $\theta: B \to G$ such that the resulting diagram is commutative. We now define

$$g: B \to \mathrm{Hom}_Z\,(R, G)$$

by $(g(b))\,(r) = \theta(rb)$, where $r \in R$ and $b \in B$. Then g is an R-homomorphism. Moreover, if $a \in A$ and $r \in R$, then

$$((g\psi)\,(a))\,(r) = \theta(r\psi(a)) = \theta\psi(ra) = \phi(ra) = (f(ra))\,(1)$$
$$= (rf(a))\,(1)$$
$$= (f(a))\,(r),$$

so $g\psi = f$ and the completed diagram Fig. 2.17 is commutative. $\square$

LEMMA 2.14 *Let E be an R-module. Then E can be embedded in the R-module $\mathrm{Hom}_Z\,(R, E)$.*

Proof. Define
$$\phi: E \to \mathrm{Hom}_Z\,(R, E)$$

by $(\phi(e))\,(r) = re$, where $r \in R$ and $e \in E$. Then ϕ is an R-homomorphism. But, if $\phi(e) = 0$ for some $e \in E$, then $e = (\phi(e))\,(1) = 0$, so that ϕ is a monomorphism. $\square$

We can now deduce Theorem 2.11. Let E be an R-module. As an Abelian group, E can be embedded in an injective Abelian group E' (Lemma 2.12). Proposition 1.25 now gives that $\mathrm{Hom}_Z\,(R, E)$ can be embedded in $\mathrm{Hom}_Z\,(R, E')$. This embedding is initially only a Z-homomorphism, but when the two modules have been given the structure of R-modules it is easy to see that it is actually an R-homomorphism. If we now use Lemma 2.14, we have a sequence of embeddings

$$E \to \mathrm{Hom}_Z(R, E) \to \mathrm{Hom}_Z(R, E').$$

But E' is injective, and Lemma 2.13 gives that $\mathrm{Hom}_Z(R, E')$ is an injective R-module. We have thus accomplished what we set out to do, namely we have embedded E in an injective module. $\square$

In view of Proposition 1.1, we may restate Theorem 2.11 as follows:

REFORMULATION OF THEOREM 2.11 *Every module has an injective extension.* It is in this form that the result is more often used.

We can employ Theorem 2.11 to provide an alternative characterization of injective modules to those already given:

THEOREM 2.15 *Let E be an R-module. Then the following statements are equivalent:*

(a) *E is injective;*

(b) *E is a direct summand of every extension of itself.*

Proof. Assume that E is injective and let E' be an extension of E. The diagram Fig. 2.19 can be completed by a homomorphism θ as shown. Let $e' \in E'$. Then $\theta(e') \in E$, so that $\theta(e') = \theta(\theta(e'))$, whence $e' - \theta(e')$ belongs to $\mathrm{Ker}\,\theta$. Thus e' belongs to $E + \mathrm{Ker}\,\theta$, so that $E' = E + \mathrm{Ker}\,\theta$. But $E \cap \mathrm{Ker}\,\theta = 0$, and Proposition 1.5 shows that $E' = E + \mathrm{Ker}\,\theta$ (d.s.). Thus E is a direct summand of E'.

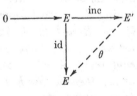

Fig. 2.19

Now assume (b). By Theorem 2.11, E has an injective extension N (say). Thus E will be a direct summand of N and so will be injective by Proposition 2.3. $\square$

2.4 Essential extensions

DEFINITION *Let the R-module E be an extension of the R-module A. Then E is said to be an 'essential extension' of A if, for every non-zero submodule E' of E, E' ∩ A ≠ 0. This is equivalent to the condition that, for every non-zero element e of E, there exists r ∈ R such that re is a non-zero element of A.*

REMARKS Every module is an essential extension of itself. Also, if A is a submodule of B which in turn is a submodule of C, then the following statements are equivalent:

(a) C is an essential extension of A;

(b) C is an essential extension of B and B is an essential extension of A.

EXAMPLE If R is a commutative domain with field of fractions K, then K, when considered as an R-module, is an essential extension of R.

REMARK Suppose we have a commutative diagram Fig. 2.20 of R-modules, where the mappings $A_1 \to A_2$ and $E_1 \to E_2$ are isomorphisms. It follows easily that, *if E_1 is an essential extension of A_1, then E_2 is an essential extension of A_2.* Combining this with Proposition 1.1, we see that, *if an R-module A has no proper essential extension (i.e. there is no R-module strictly containing A which is an essential extension of A), then neither has any R-module which is isomorphic to A.*

Fig. 2.20

PROPOSITION 2.16 *Let $\{A_\lambda\}_{\lambda \in \Lambda}$ be a family of R-modules and suppose that, for each λ, A_λ has an essential extension E_λ. Then $\underset{\lambda \in \Lambda}{\oplus} E_\lambda$ is an essential extension of $\underset{\lambda \in \Lambda}{\oplus} A_\lambda$.*

Proof. Put $A = \underset{\lambda \in \Lambda}{\oplus} A_\lambda$, $E = \underset{\lambda \in \Lambda}{\oplus} E_\lambda$, and consider a non-zero element e of E. Then e has only a finite number of non-zero components; let these be $e_i \in E_{\lambda_i}$ for $i = 1, 2, \ldots, n$. Because E_{λ_1} is an essential extension of A_{λ_1}, there exists $r_1 \in R$ such that $r_1 e_1$ is a non-zero element of A_{λ_1}. Consider $r_1 e$. This is a non-zero element of E. It may be that the only non-zero component is $r_1 e_1$. If so, then $r_1 e \in A$ and the proof is complete. Otherwise, let $r_1 e_p$ be the first non-zero element in the sequence $r_1 e_2, r_1 e_3, \ldots, r_1 e_n$. Because E_{λ_p} is an essential extension of A_{λ_p}, there exists $r_p \in R$ such that $r_p r_1 e_p$ is a non-zero element of A_{λ_p}. Note also that $r_p r_1 e_1 \in A_{\lambda_1}$. It should now be clear how the argument proceeds. At the final stage, there is an element r of R such that re is a non-zero element of A. This shows that E is an essential extension of A. $\square$

We now give a characterization of injective modules in terms of essential extensions.

THEOREM 2.17 *Let E be an R-module. Then the following statements are equivalent:*

(a) *E is injective;*

(b) *E has no proper essential extension.*

Proof. Suppose that E is injective and let E' be a proper extension of E. By Theorem 2.15, E is a direct summand of E', so there exists a non-zero submodule F of E' such that $E' = E + F$ (d.s.). But then $E \cap F = 0$, so that E' is not an essential extension of E. This proves (b).

Now assume (b) and let F be an extension of E. By Theorem 2.15, (a) will follow if we show that E is a direct summand of F. It may be assumed that F is a proper extension of E. Denote by Ω the collection of all non-zero submodules X of F such that $E \cap X = 0$. By (b), Ω is not empty. When partially ordered by inclusion, Ω is an inductive system and so by Zorn's Lemma has a maximal member X_0 (say). We shall show that $F = E + X_0$, from which it will follow that $F = E + X_0$ (d.s.).

Using Proposition 1.10 Corollary 3 we have that

$$F/X_0 \supseteq (E + X_0)/X_0 \approx E/(E \cap X_0) \approx E.$$

Suppose that $F \neq E + X_0$. Then, by Proposition 1.9,

$$F/X_0 \supset (E + X_0)/X_0$$

(the inclusion being strict). But E has no proper essential extension, so neither has $(E + X_0)/X_0$. It follows by means of Proposition 1.9 again that there exists a submodule Y of F such that $Y \supset X_0$ and $(Y/X_0) \cap ((E + X_0)/X_0) = 0$. But now (1.5.2) gives that $Y \cap (E + X_0) = X_0$, so that

$$Y \cap E \subseteq Y \cap (E + X_0) = X_0.$$

Thus $Y \cap E \subseteq E \cap X_0 = 0$, i.e. $Y \in \Omega$. This contradicts the maximality of X_0 and (a) is now established. $\square$

PROPOSITION 2.18 *Let A be an R-module, let E be an essential extension of A and let N be an injective extension of A. Then the inclusion mapping of A into N can be extended to an embedding of E in N.*

Proof. Consider Fig. 2.21. Because N is injective, there is an R-homomorphism $g: E \to N$ such that the resulting diagram is commutative, i.e. g extends the inclusion mapping $A \to N$. It follows that $A \cap \operatorname{Ker} g = 0$. But E is an essential extension of A, so that $\operatorname{Ker} g = 0$ and g is an embedding of E in N. $\square$

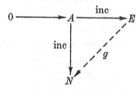

Fig. 2.21

DEFINITION *Let the R-module E be an extension of the R-module A. Then E is said to be a 'maximal essential extension' of A if*

(i) *E is an essential extension of A, and*

(ii) *whenever E' is a proper extension of E, then E' is not an essential extension of A.*

In view of a remark at the beginning of this section, the second condition in this definition is equivalent to the assertion that E has no proper essential extension.

DEFINITION *Let the R-module N be an extension of the R-module A. Then N is said to be a 'minimal injective extension' of A if*

(i) *N is injective, and*

(ii) *whenever N' is a proper submodule of N which contains A, then N' is not injective.*

PROPOSITION 2.19 *Let A be an R-module and let N be an injective extension of A. Then N has a submodule E which is a maximal essential extension of A.*

Proof. Denote by Ω the collection of all essential extensions of A which are submodules of N. Then Ω is not empty because A belongs to Ω. When Ω is partially ordered by inclusion, it is an inductive system and so by Zorn's Lemma has a maximal member E (say). We assert that E is actually a maximal essential extension of A. Let E' be an essential extension of E. By Proposition 2.18, there is a monomorphism $\theta\colon E' \to N$ which extends the inclusion mapping $E \to N$. Thus $\theta(E')$ is a member of Ω containing E, so that $\theta(E') = E$, i.e. $E' = E$.□

PROPOSITION 2.20 *Let A be an R-module and let E be an extension of A. Then the following statements are equivalent:*

(a) *E is an essential injective extension of A;*

(b) *E is a maximal essential extension of A;*

(c) *E is a minimal injective extension of A.*

Proof. That (a) and (b) are equivalent is obvious from Theorem 2.17. Now assume (b). Then E must be injective by Theorem 2.17. Let E' be an injective extension of A contained in E. Then E is an essential extension of E', so $E = E'$ by Theorem 2.17 applied to E'. Hence E is a minimal injective extension of A, which proves (c). Now assume (c). By Proposition 2.19, E has a submodule E'' which is a maximal essential extension of A and so injective. It follows that $E'' = E$ and (b) is established.□

THEOREM 2.21 *Let A be an R-module. Then there exists an R-module E satisfying the following equivalent conditions:*

(a) *E is an essential injective extension of A;*

(b) *E is a maximal essential extension of A;*

(c) *E is a minimal injective extension of A.*

Fig. 2.22

Moreover, if E_1, E_2 are both essential injective extensions of A, then there is an isomorphism $\theta\colon E_1 \to E_2$ such that Fig. 2.22 is commutative.

Proof. Theorem 2.11 and Propositions 2.19 and 2.20 give the existence of a module E with the required properties. Now let E_1, E_2 be two such modules. By Proposition 2.18, the inclusion mapping $A \to E_2$ can be extended to a monomorphism $\theta\colon E_1 \to E_2$. Then $\theta(E_1)$ is an injective extension of A contained in E_2, so that $\theta(E_1) = E_2$, i.e. θ is an isomorphism. This is precisely the situation of Fig. 2.22.□

REMARK Suppose that the R-module E_1 is an essential injective extension of the R-module A, and suppose that E_2 is another extension of A such that there is an isomorphism $\theta: E_1 \to E_2$ making Fig. 2.22 commutative. Then E_2 is also an essential injective extension of A. Thus the uniqueness clause in Theorem 2.21 cannot be strengthened.

DEFINITION *Let A be an R-module. An R-module E satisfying the conditions of Theorem 2.21 is called an 'injective envelope' of A;† we use the symbol $E(A)$ to denote an injective envelope of A.*

REMARK An R-module A is injective if and only if $E(A) = A$. Also, if A is an R-module and B is a submodule of $E(A)$ which contains A, then $E(A)$ is also an injective envelope of B.

EXAMPLE Let R be a commutative domain. The field of fractions of R, when considered as an R-module, is an injective envelope of R.

REMARK Let A_1, A_2 be R-modules, and suppose that we have an isomorphism $\theta: A_1 \to A_2$. Consider injective envelopes $E(A_1)$, $E(A_2)$ of A_1, A_2 respectively. By Proposition 1.1, A_2 has an extension E' such that $E' \approx E(A_1)$, where this isomorphism

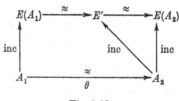

Fig. 2.23

extends θ. Then E' is an injective envelope of A_2, and we can insert the isomorphism $E' \approx E(A_2)$ in Fig. 2.23, the whole diagram being commutative. It follows that $E(A_1) \approx E(A_2)$ under an isomorphism which extends θ. In other words, *an isomorphism $A_1 \approx A_2$ between R-modules can be extended to an isomorphism $E(A_1) \approx E(A_2)$ between their respective injective envelopes.*

† The term 'injective hull' is used by some authors in place of 'injective envelope'.

PROPOSITION 2.22   *Let A be a submodule of an R-module B and let $E(B)$ be an injective envelope of B. Then A has an injective envelope $E(A)$ which is a submodule of $E(B)$, and $E(A)$ is a direct summand of $E(B)$. Moreover, if B is an essential extension of A, then $E(B) = E(A)$.*

Proof. The first part follows from Proposition 2.19 and Theorem 2.15. If B is an essential extension of A, then $E(B)$ is an essential extension of A, whence $E(B) = E(A)$.□

PROPOSITION 2.23   *Let $A_1, A_2, ..., A_n$ be R-modules. Then $\bigoplus_{i=1}^{n} E(A_i)$ is an injective envelope of $\bigoplus_{i=1}^{n} A_i$, i.e.*

$$E(A_1 \oplus A_2 \oplus ... \oplus A_n) = E(A_1) \oplus E(A_2) \oplus ... \oplus E(A_n).$$

Proof. Because there are only finitely many A_i, the module $\bigoplus_{i=1}^{n} E(A_i)$ is injective (Proposition 2.3). By Proposition 2.16, it is also an essential extension of $\bigoplus_{i=1}^{n} A_i$.□

REMARK   We shall see in Proposition 4.2 that, when R is a left Noetherian ring, the analogous result is true for an arbitrary collection of R-modules A_i, and not just for a finite number.

PROPOSITION 2.24   *Let A be an R-module and let a be a nonzero element of A. Then there is a simple module S and a homomorphism $\phi: A \to E(S)$ such that $\phi(a) \neq 0$.*

Proof. Consider the left ideal $0:a$, consisting of all elements r of R such that $ra = 0$. This is a proper left ideal because $a \neq 0$, so it is contained in a maximal left ideal M of R. We define a mapping $\phi': Ra \to R/M$ by

$$\phi'(ra) = r + M.$$

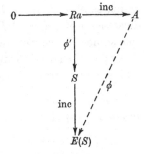

Fig. 2.24

Note that ϕ' is well-defined, because if $ra = 0$ then $r \in 0:a \subseteq M$ and $r + M = 0_{R/M}$. Further, ϕ' is an R-homomorphism and $\phi'(a) = 1 + M \neq 0_{R/M}$. We put $S = R/M$, so that R/M is a simple R-module, and consider Fig. 2.24. Since $E(S)$ is injective, there exists a homomorphism $\phi: A \to E(S)$ such that $\phi(a) = \phi'(a) \neq 0$.□

DEFINITION *An R-module E is said to be an 'injective cogenerator' of R if* (i) *E is injective, and* (ii) *for every R-module A and every non-zero element a of A, there is an R-homomorphism* $\phi: A \to E$ *such that* $\phi(a) \neq 0$.

Consider all isomorphism classes of simple R-modules, and let $\{S_i\}_{i \in I}$ be a family of representatives, one from each isomorphism class. By Proposition 1.12, we could take $S_i = R/M_i$, where M_i is a maximal left ideal of R, so this amounts to considering a family of distinct maximal left ideals of R. Indeed, (1.6.1) shows that, when R is commutative, this family consists of all the maximal ideals of R. It follows from Proposition 2.24 that $\prod_{i \in I} E(S_i)$ is an injective cogenerator of R; so is $E(\bigoplus_{i \in I} E(S_i))$.

PROPOSITION 2.25 *Let E be an injective cogenerator of R. Then every R-module can be embedded in a direct product of copies of E.*

Proof. It is sufficient to consider a non-zero R-module A. Let a be a non-zero element of A. There is an R-homomorphism $\phi_a: A \to E$ such that $\phi_a(a) \neq 0$. We define

$$\phi: A \to \prod_{\substack{a \in A \\ a \neq 0}} E$$

by $\phi(x) = \{\phi_a(x)\}$ for $x \in A$. Then ϕ is a monomorphism. $\square$

Proposition 2.25 may be regarded as a dual of Proposition 1.4, which says that every R-module is a homomorphic image of a direct sum of copies of R. An injective cogenerator of R plays a role somewhat complementary to that of R itself.

A special case of Proposition 2.25 will be useful in Chapter 5, and is contained in the following corollary.

COROLLARY *Every R-module can be embedded in a direct product of injective envelopes of simple modules.* $\square$

PROPOSITION 2.26 *Let I be a proper left ideal of R and let* $r \in \operatorname{Ann}_R E(R/I)$. *Then there is an element* $s \in R$, $s \notin I$, *such that* $sr = 0$.

Proof. The argument is very similar to that of Proposition 2.24. Suppose that there is no element $s \in R$, $s \notin I$ such that $sr = 0$. Then $0:r \subseteq I$ and we can define an R-homomorphism

$\phi'\colon Rr \to R/I$ by $\phi'(r'r) = r' + I$, where $r' \in R$. In the manner of Proposition 2.24, this can be extended to an R-homomorphism $\phi\colon R \to E(R/I)$. We note that

$$r\phi(1) = \phi(r) = \phi'(r) = 1 + I \neq 0.$$

But $\phi(1) \in E(R/I)$ and $r \in \operatorname{Ann}_R E(R/I)$, so that $r\phi(1) = 0$ by the very definition of the annihilator. This provides a contradiction. $\square$

COROLLARY 1   *Let R be a domain and let I be a proper left ideal of R. Then $\operatorname{Ann}_R E(R/I) = 0$.* $\square$

COROLLARY 2   *Suppose that R has a unique maximal left ideal M. Then $\operatorname{Ann}_R E(R/M) = 0$.*

Proof. Let $r \in \operatorname{Ann}_R E(R/M)$. Then there exists $s \in R$, $s \notin M$, such that $sr = 0$. But the left ideal generated by s must be the whole ring, so there exists $t \in R$ such that $ts = 1$. It follows that $r = 0$. $\square$

PROPOSITION 2.27   *Let I be a two-sided ideal of R and let E be an injective R-module. Then $0\colon_E I$ is injective as an (R/I)-module. Moreover, if E is the injective envelope of an R-module A, then $0\colon_E I$ is an injective envelope of $A \cap (0\colon_E I)$ considered as an (R/I)-module.*

Proof. Consider the diagram Fig. 2.25 where B and C are (R/I)-modules, f and g are (R/I)-homomorphisms and g is a monomorphism. By Proposition 1.13, B, C and $0\colon_E I$ can be considered as R-modules, when f and g are R-homomorphisms. Because E is injective, the diagram can be completed by an R-homomorphism h as shown. Let $r \in I$, $c \in C$. Then

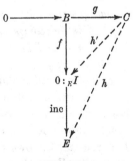

Fig. 2.25

$rh(c) = h(rc) = 0$, so that $h(C) \subseteq 0\colon_E I$.

We can thus fill in the R-homomorphism h' as shown so that the diagram is commutative; h and h' are to all intents and purposes the same. Moreover h' is an (R/I)-homomorphism. This shows that $0\colon_E I$ is injective as an (R/I)-module.

We turn now to the second statement of the proposition. Suppose that $E = E(A)$, where A is an R-module. Let X be any non-zero (R/I)-submodule of $0:_E I$. Then X is also an R-submodule of $0:_E I$, and $X \cap (A \cap (0:_E I)) = X \cap A \neq 0$ because E is an essential extension of A. It follows that $0:_E I$ is an essential extension of $A \cap (0:_E I)$ and so is the injective envelope of $A \cap (0:_E I)$. $\square$

2.5 Indecomposable injective modules

DEFINITION *An R-module E is said to be 'indecomposable' if (a) $E \neq 0$ and (b) the only direct summands of E are 0 and E itself. If a module is indecomposable, then so is every module which is isomorphic to it.*

DEFINITION *Let E be an R-module and let M be a submodule of E. Then M is said to be an 'irreducible submodule' of E if (a) $M \neq E$, and (b) there do not exist submodules M_1, M_2 of E such that $M_1 \supset M$, $M_2 \supset M$ and $M_1 \cap M_2 = M$ (the inclusions being strict). A left ideal of a ring is said to be an 'irreducible left ideal' if, when the ring is regarded as a left module over itself, the left ideal is an irreducible submodule.*

PROPOSITION 2.28 *Let E be an injective R-module. Then the following statements are equivalent:*

  (a) E is indecomposable;

  (b) $E \neq 0$ and is an injective envelope of every non-zero submodule of itself;

  (c) the zero submodule of E is irreducible.

Proof. Suppose first that E is indecomposable. Then $E \neq 0$. Let M be a non-zero submodule of E. By Proposition 2.19, E has a submodule E' which is an injective envelope of M, and Theorem 2.15 shows that E' is a direct summand of E. Since $E' \neq 0$ it follows that $E' = E$. This proves (b).

Now assume (b) and suppose that M_1, M_2 are submodules of E such that $M_1 \cap M_2 = 0$. Suppose that $M_1 \neq 0$. Then $E = E(M_1)$ so, in particular, E is an essential extension of M_1. It follows that $M_2 = 0$. This proves (c).

Now assume (c). Then $E \neq 0$. Suppose that there exist non-

zero submodules E_1, E_2 of E such that $E = E_1 + E_2$ (d.s.). By
Proposition 1.5, $E_1 \cap E_2 = 0$, which contradicts (c). Hence E is
indecomposable. $\square$

COROLLARY 1   *Let M be an R-module. Then*
   *(a) $E(M)$ is indecomposable if and only if the zero submodule of
M is irreducible;*
   *(b) a submodule K of M is irreducible if and only if $E(M/K)$
is indecomposable.*
   Proof. (a) Suppose that $E(M)$ is indecomposable. By the
proposition, the zero submodule of $E(M)$ is irreducible, from
which it follows that the zero submodule of M is irreducible.
Conversely, suppose that the zero submodule of M is irreducible
and let E_1, E_2 be submodules of $E(M)$ such that $E_1 \cap E_2 = 0$. Now
$E(M)$ is an essential extension of M, so if $E_1 \neq 0$ then $E_1 \cap M \neq 0$.
Similarly, if $E_2 \neq 0$ then $E_2 \cap M \neq 0$. But $(E_1 \cap M) \cap (E_2 \cap M) = 0$
and $E_1 \cap M$, $E_2 \cap M$ are submodules of M. It follows that one of
E_1, E_2 must be zero, so that the zero submodule of $E(M)$ is irreducible. This in turn implies that $E(M)$ is indecomposable.
   (b) Let K_1, K_2 be submodules of M containing K. Then
$K_1 \cap K_2 = K$ if and only if $(K_1/K) \cap (K_2/K) = 0$, so that the submodule K of M is irreducible if and only if the zero submodule of
M/K is irreducible. With this observation, (b) follows from (a). $\square$

COROLLARY 2   *If S is a simple R-module, then $E(S)$ is
indecomposable.*
   Proof. This follows from Corollary 1 because the zero submodule of a simple module is irreducible. $\square$

COROLLARY 3   *Let E be an indecomposable injective R-module
and let e be a non-zero element of E. Then $E \approx E(R/(0:e))$ and $0:e$
is an irreducible left ideal of R.*
   Proof. By $(1.6.2)$, $Re \approx R/(0:e)$ and the proposition gives that
$E = E(Re) \approx E(R/(0:e))$. Then $E(R/(0:e))$ is indecomposable and
Corollary 1 shows that $0:e$ is an irreducible left ideal of R. $\square$

COROLLARY 4   *Let E be an R-module. Then E is an indecomposable injective R-module if and only if there is an irreducible
left ideal I of R such that $E \approx E(R/I)$.*
   Proof. This follows from Corollaries 1 and 3. $\square$

4   SVI

COROLLARY 5   *Let E be an indecomposable injective R-module and suppose that either E or R is Artinian. Then $E = E(S)$ for some simple R-module S.*

Proof. This follows from Propositions 1.20, 1.21 and 2.28. ☐

EXAMPLE   Let R be a commutative domain with quotient field K. Then K, considered as an R-module, is an injective envelope of R. Also, K is an indecomposable injective R-module because the zero ideal of R is irreducible.

When the ring R is commutative and Noetherian, we can give a complete description of the indecomposable injective R-modules in terms of the prime ideals of R. This will enable us to describe completely the indecomposable injective Abelian groups. *For the rest of this section, R will denote a commutative ring.*

DEFINITION   *Let P be an ideal of R. Then P is said to be a 'prime ideal' of R if (a) P is proper, and (b) whenever*

$$\alpha\beta \in P \quad (\alpha, \beta \in R),$$

then either $\alpha \in P$ or $\beta \in P$.

REMARKS

1. Let P be a prime ideal of R and let $r_1, r_2, ..., r_n$ $(n \geqslant 1)$ be elements of R. If $r_1 r_2 ... r_n \in P$, then $r_i \in P$ for some i. In particular, if $r \in R$ and $r^n \in P$, then $r \in P$.

2. Let P be a prime ideal of R and let A, B be ideals of R such that $AB \subseteq P$. Then either $A \subseteq P$ or $B \subseteq P$.

3. If P is a prime ideal of R, then the residue class ring R/P is an integral domain.

LEMMA 2.29   *Let P be a prime ideal of R. Then P is irreducible and $E(R/P)$ is indecomposable.*

Proof. Suppose that $P = A \cap B$, where A and B are ideals of R. Then $AB \subseteq A \cap B = P$, so either $A \subseteq P$ or $B \subseteq P$, i.e. either $P = A$ or $P = B$. Hence P is irreducible. It follows from Proposition 2.28 Corollary 1 that $E(R/P)$ is indecomposable. ☐

LEMMA 2.30   *Let M be a non-zero Noetherian R-module. Then there exists an element m of M such that $0:m$ is a prime ideal of R.*

Proof. Denote by Ω the collection of all ideals of R of the form $0:x$, where x is a non-zero element of M. Certainly Ω is not empty. We contend that Ω has a maximal member. For suppose not. Then there is an infinite strictly ascending chain

$$(0:x_1) \subset (0:x_2) \subset (0:x_3) \subset \dots$$

of elements of Ω, and corresponding to this we obtain the infinite strictly ascending chain

$$0 \subset (0:x_2)/(0:x_1) \subset (0:x_3)/(0:x_1) \subset \dots$$

of submodules of the R-module $R/(0:x_1)$. Now $R/(0:x_1) \approx Rx_1$, which is a submodule of M, so that $R/(0:x_1)$ is Noetherian and so cannot possess an infinite strictly ascending chain of submodules. This provides the required contradiction.

Let $0:m = P$ (say) be a maximal member of Ω. We contend that P is prime. Certainly P is proper, so let α, β be elements of R such that $\alpha\beta \in P$. Assume that $\beta \notin P$. Then $\alpha\beta m = 0$ but $\beta m \neq 0$. Now $0:\beta m \in \Omega$ and $0:\beta m \supseteq 0:m$, so that

$$0:\beta m = 0:m = P.$$

But $\alpha \in 0:\beta m$, so that $\alpha \in P$. $\square$

COROLLARY 1 *Let M be a non-zero Noetherian R-module. Then there is a prime ideal P of R such that M has a submodule isomorphic to R/P.*

Proof. By the lemma, there exists $m \in M$ such that $0:m$ is a prime ideal of R, and $Rm \approx R/(0:m)$. $\square$

COROLLARY 2 *Every maximal ideal of R is prime.*

Proof. Let M be a maximal ideal of R. Then R/M is a simple R-module, so, from Corollary 1, there exists a prime ideal P of R such that $R/P \approx R/M$. It follows from (1.6.1) that $P = M$ and M is prime. $\square$

Let P be a prime ideal of R. It is an easy consequence of the definition that, if ξ is any non-zero element of the R-module R/P, then $0:\xi = P$. Now let η be a non-zero element of $E(R/P)$. There exists an element r of R such that $r\eta$ is a non-zero element of R/P, whence $0:r\eta = P$. But $0:\eta \subseteq 0:r\eta$. This proves the following result:

LEMMA 2.31   *Let P be a prime ideal of R. Then the collection of all annihilators of non-zero elements of the R-module $E(R/P)$ has a unique maximal member, namely P itself.* □

COROLLARY   *Let P_1, P_2 be prime ideals of R and suppose that $E(R/P_1) \approx E(R/P_2)$. Then $P_1 = P_2$.* □

DEFINITION   *Let P be an ideal of R. Then P is said to be 'N-prime' if (a) P is prime, and (b) R/P is a Noetherian ring.*

For example, every maximal ideal M of R is N-prime; it is prime by Lemma 2.30 Corollary 2 and R/M is a field.

DEFINITION   *Let E be an R-module. Then E is said to be 'N-injective' if (a) E is injective, and (b) whenever $E \neq 0$, then E possesses a non-zero Noetherian submodule.*

If an R-module is N-injective, then so is every R-module which is isomorphic to it.

If R is a Noetherian ring, there is no difference between its prime and its N-prime ideals. Nor is there any difference between the injective R-modules and the N-injective R-modules. For a non-zero R-module certainly has a non-zero finitely generated submodule, and this will be Noetherian when R is a Noetherian ring by Proposition 1.19.

THEOREM 2.32
(i) *If P is an N-prime ideal of R, then $E(R/P)$ is an indecomposable N-injective R-module.*

(ii) *If E is an indecomposable N-injective R-module, then there is a unique N-prime ideal P of R such that $E \approx E(R/P)$.*

Proof. (i) Let P be an N-prime ideal of R. Then, by Lemma 2.29, $E(R/P)$ is indecomposable. Since $E(R/P)$ has R/P as a non-zero Noetherian submodule, it follows that $E(R/P)$ is N-injective.

(ii) Let E be an indecomposable N-injective R-module. By Lemma 2.30 Corollary 1, E has a Noetherian submodule M which is isomorphic to R/P for some prime ideal P. It follows that $E = E(M) \approx E(R/P)$. Also, P is N-prime. The uniqueness of P follows from Lemma 2.31 Corollary. □

Theorem 2.32 gives a complete description of the indecomposable injective R-modules when R is a Noetherian ring.

CorOLLARY   *Let R be a commutative Noetherian ring and let E be an R-module. Then the following statements are equivalent:*
  (a) E is an indecomposable injective R-module;
  (b) $E \approx E(R/P)$ for some prime ideal P of R. $\square$

As an illustration of this last result, we shall determine all the indecomposable injective Abelian groups. The ring Z of integers is a principal ideal domain, and so is Noetherian. Its prime ideals are (a) the zero ideal and (b) those ideals of the form Zp, where p is a prime number. The zero ideal gives rise to the indecomposable injective Abelian group $E(Z)$, which we know from the example immediately following the definition of injective envelopes in Section 2.4 is Q, the additive group of rational numbers. Now let p be a prime number, and denote by Z_p the set of rational numbers of the form m/n, where m and n are integers and n is not divisible by p. Then Z_p is a subgroup of Q. We write

$$Z(p^\infty) = Q/Z_p$$

and assert that $E(Z/Zp) \approx Z(p^\infty)$.

First, Q is a divisible Abelian group, and Lemma 2.4 shows that $Z(p^\infty)$ is also divisible, and so injective by Theorem 2.8. Denote by W the subgroup of $Z(p^\infty)$ generated by the single element $(1/p) + Z_p$. Then $Z(p^\infty)$ is an essential extension of W. For consider a non-zero element $(a/b) + Z_p$ $(a, b \in Z, b \neq 0)$ of $Z(p^\infty)$. We shall suppose that the fraction a/b is in lowest terms, so that $p|b$ but $p \nmid a$. Write $b = p^k b'$, where $b' \in Z$ and $p \nmid b'$. Then

$$p^{k-1}b'((a/b) + Z_p) = (a/p) + Z_p,$$

which is a non-zero element of W. Thus $Z(p^\infty) = E(W)$. But

$$Z/Zp \approx W,$$

the element $n + Zp$ $(n \in Z)$ corresponding to $(n/p) + Z_p$. Hence $E(Z/Zp) \approx E(W) = Z(p^\infty)$.
  To sum up,

$$E(Z) = Q \quad \text{and} \quad E(Z/Zp) \approx Z(p^\infty).$$

Lemma 2.31 Corollary shows that Q and the $Z(p^\infty)$ are non-isomorphic. We state these conclusions in a theorem

THEOREM 2.33   *The Abelian groups Q and $Z(p^\infty)$, where p is a prime number, are non-isomorphic, indecomposable and injective. Further, every indecomposable injective Abelian group is isomorphic to one of these.* $\square$

Exercises on Chapter 2

2.1   Let E be an R-module. A system of equations in x of the form $r_i x = e_i$ $(i \in I)$, where $r_i \in R$ and $e_i \in E$, is said to be *compatible* if, whenever $\sum_{i \in I} s_i r_i = 0$, where $s_i \in R$ for each $i \in I$ and all but a finite number of the s_i are zero, then $\sum_{i \in I} s_i e_i = 0$. Show that E is injective if and only if every such compatible system of equations has a solution in E.

2.2   Let A be an R-module and let E be an injective R-module. Let $a_1, a_2, \dots, a_n \in A$ $(n \geqslant 1)$ and let $e \in E$. Show that
$$(0:a_1) \cap (0:a_2) \cap \dots \cap (0:a_n) \subseteq (0:e)$$
if and only if there exist $f_1, f_2, \dots, f_n \in \mathrm{Hom}_R (A, E)$ such that
$$e = f_1(a_1) + f_2(a_2) + \dots + f_n(a_n).$$
[*Hint:* Consider the cyclic R-module
$$R/((0:a_1) \cap (0:a_2) \cap \dots \cap (0:a_n)).]$$

2.3   Let R be a commutative domain with quotient field K and let I be a finitely generated submodule of K, where K is considered as an R-module. Show that I is a fractional ideal of R.

2.4   Let R be a commutative domain and let E be a non-zero torsion-free divisible R-module. Show that the ring of endomorphisms of E has a subring isomorphic to the field of fractions of R.

2.5   Let $\{A_i\}_{i \in I}$ and $\{B_i\}_{i \in I}$ be families of submodules of an R-module M such that, for each i, A_i is an essential extension of B_i. Suppose that the sum $\sum B_i$ is direct. Prove that the sum $\sum_{i \in I} A_i$ is direct.

2.6   Let R be a commutative domain with quotient field K and let I be a non-zero fractional ideal of R. Show that K, considered as an R-module, is an injective envelope of I.

2.7   A commutative ring R is said to be a *valuation ring* if, for every pair of elements a, b of R, either $a \in Rb$ or $b \in Ra$. Let R

be a commutative valuation domain with quotient field K and let A be a proper ideal of R. Show that K/A is an essential extension of R/A.

2.8 Let M be an R-module. We define the *tertiary radical*, $t(M)$, of M to be the set of all elements r of R for which there is a submodule M' of M such that M is an essential extension of M' and $rM' = 0$. Show that

   (i) $t(M)$ is a two-sided ideal of R;

   (ii) if N is a submodule of M, then $t(N) \supseteq t(M)$;

   (iii) if M is an essential extension of N, then $t(N) = t(M)$.

2.9 A domain is called a *left Ore domain* if, when regarded as a left module over itself, its zero submodule is irreducible. Prove the following:

   (i) a commutative domain is an Ore domain;

   (ii) the domain R is a left Ore domain if and only if, given non-zero elements a, b of R, there exist non-zero elements r, s of R such that $ra = sb$;

   (iii) every injective envelope of a left Ore domain, regarded as a left module over itself, is torsion-free.

2.10 Generalize Proposition 2.7 to left Ore domains. (See Exercise 2.9.)

2.11 A ring R is said to be *left hereditary* if every homomorphic image of every injective left R-module is injective. Establish the equivalence of the following statements:

   (a) the ring R is left hereditary;

   (b) wherever M_1, M_2 are injective submodules of a left R-module M, then $M_1 + M_2$ is injective. [*Hint:* To show that (b) implies (a), consider a submodule N of an injective R-module M. Put $H = M \oplus M$ and $K = \{(n, n) : n \in N\}$. Show that H/K is injective. Then show that M/N is isomorphic to a certain factor module of H/K.]

2.12 Let E be an injective cogenerator of R. Show that the sequence
$$A \to B \to C$$
of R-modules and R-homomorphisms is exact if and only if the sequence
$$\operatorname{Hom}_R(C, E) \to \operatorname{Hom}_R(B, E) \to \operatorname{Hom}_R(A, E)$$
is exact.

2.13 Let R be a commutative ring and let I be an ideal of R. For a maximal ideal M of R, put

$$I_M = 0:(0:_{E(R/M)}I).$$

Prove that

(i) $I \subseteq I_M$ for every maximal ideal M of R;

(ii) $I_M = R$ if and only if $I \nsubseteq M$;

(iii) $I = \bigcap_M I_M$, where M ranges over all the maximal ideals of R which contain I. [*Hint*: In (iii), use Proposition 2.24.]

2.14 A ring R is said to be *regular* if, for every $a \in R$, there exists $b \in R$ such that $a = aba$. Let R be a commutative ring. Show that R is regular if and only if every simple R-module is injective.

2.15 Suppose that every proper left ideal I of the ring R can be written in the form $I = J \cap K$, where J and K are left ideals of R, $K \supset I$ and J is irreducible. Show that every non-zero injective R-module has an indecomposable injective submodule.

2.16 Let I be a proper left ideal of R such that $E(R/I)$ has an indecomposable injective submodule. Show that there exist left ideals J and K of R such that $I = J \cap K$, $K \supset I$ and J is irreducible. (This provides a converse to Exercise 2.15.)

2.17 Let R be the ring of all sequences of elements of some field F, addition and multiplication being defined componentwise. Let I be the ideal consisting of all sequences with only finitely many non-zero terms. Show that

(i) every element of R is of the form ue, where u is a unit and e is idempotent (i.e. $e^2 = e$);

(ii) every prime ideal of R is maximal;

(iii) every irreducible ideal of R is maximal;

(iv) I cannot be expressed in the form $I = J \cap K$, where $K \supset I$ and J is irreducible;

(v) $E(R/I)$ has no indecomposable injective submodules. [*Hint:* Use Exercise 2.16];

(vi) R is a regular ring (see Exercise 2.14).

Notes on Chapter 2

Divisible groups, and then injective modules, were introduced by R. Baer, B. Eckmann and A. Schopf. The dual notion of a projective cover does not always exist for modules. H. Bass [2] has characterized those rings for which every module has a projective cover. For injective objects in a general Abelian category, see [7], [9] and [21]. A. Grothendieck gave the right axiom to ensure that every object in an Abelian category can be embedded in an injective object. This is the so-called axiom $AB5$.

Injectives play an important role in categories other than the category of modules. In the category of Banach spaces, the field of real numbers is injective; this is the Hahn–Banach theorem. In the category of Boolean algebras, a complete Boolean algebra is injective (see [12]). In the category of normal topological spaces, the closed interval $[0, 1]$ is injective; this is Tietze's theorem (see [14]). In the category of partially ordered sets, the injective envelope is the MacNeille completion (B. Banaschewski and G. Bruns [1]). Some of the results on injectives in the category of modules can be carried over into these categories; the most notable one is the result that a direct product of injectives is injective (Proposition 2.2).

The observation that the injective envelope of a commutative domain (or Ore domain) is its field of fractions (quotients) has led to the notion of the generalized ring of quotients of an arbitrary ring. Many people have contributed to this. We refer the reader to [16].

It was E. Matlis who realized that indecomposable injective modules can play the role of prime ideals (see Section 2.5 and Chapter 4).

3. *Injective modules and semi-simplicity*

Throughout this chapter, the symbol R will continue to denote a ring with an identity element. It is not assumed that R is commutative. The term 'R-module' will continue to mean '*left R-module*'.

3.1 Semi-simple rings and modules

LEMMA 3.1 *Let M be an R-module such that every submodule of M is a direct summand of M and let M' be a submodule of M. Then every submodule of M' is a direct summand of M'.*

Proof. Let A be a submodule of M'. Then A is also a submodule of M, so there is a submodule B of M such that $M = A + B$ (d.s.). It follows by Proposition 1.2 that $M' = A + (M' \cap B)$, and this sum is direct since $A \cap (M' \cap B) = A \cap B = 0$ (see Proposition 1.5). $\square$

PROPOSITION 3.2 *Let M be an R-module. Then the following statements are equivalent:*

(a) *M has a family $\{S_i\}_{i \in I}$ of simple submodules such that $M = \sum_{i \in I} S_i$ (d.s.);*

(b) *M has a family of simple submodules whose sum is M itself;*

(c) *every submodule of M is a direct summand of M.*

REMARK We shall provide a proof which uses the results of Chapter 2. For a proof of a more elementary nature, we refer the reader to, for example, H. Cartan and S. Eilenberg, *Homological Algebra* (Princeton University Press, 1956), Chapter 1, Proposition 4.1. Recall that a zero module is regarded as the (direct) sum of the empty family of simple submodules.

Proof. We may suppose that $M \neq 0$. Clearly (a) implies (b). We now prove that (b) implies (c). Suppose that $M = \sum_{i \in I} S_i$, where the S_i are simple modules. We first show that, if N is a submodule of M such that M is an essential extension of N, then $M = N$. Suppose indeed that M is an essential extension of the submodule N. Then, for each $i \in I$, $S_i \cap N \neq 0$, whence $S_i \cap N = S_i$ because S_i is simple. It follows that $S_i \subseteq N$ for each i, whence $M = N$, as asserted.

Now let M' be a submodule of M. We wish to show that M' is a direct summand of M. We may suppose that $M' \neq M$. By Proposition 2.22, there are injective envelopes $E(M)$ and $E(M')$ such that $E(M')$ is a direct summand of $E(M)$, i.e. there is a submodule F of $E(M)$ such that $E(M) = E(M') + F$ (d.s.). Now F is an essential extension of $M \cap F$. For consider a non-zero submodule T of F. Then T is also a submodule of $E(M)$, so that $T \cap M \cap F = T \cap M \neq 0$, as required. It now follows from Proposition 2.16 that $E(M) = E(M') + F$ (d.s.) is an essential extension of $M' + (M \cap F)$, whence M is an essential extension of $M' + (M \cap F)$. Hence $M = M' + (M \cap F)$ (d.s.) by what we have shown above, which shows that M' is a direct summand of M, as required.

Now assume (c) and let x be a non-zero element of M. By $(1.6.2)$, $Rx \approx R/(0:x)$, and $0:x$ is a proper left ideal of R. Thus $0:x$ is contained in a maximal left ideal Π of R, and the submodule $\Pi/(0:x)$ of $R/(0:x)$ transfers across to a maximal submodule N of Rx. Thus Rx/N is a simple module. But, by Lemma 3.1, N is a direct summand of Rx, so there exists a submodule S such that $Rx = N + S$ (d.s.). Now, by Proposition 1.11, $S \approx Rx/N$ and so is simple. But x was an arbitrary non-zero element of M, so every non-zero submodule of M has a simple submodule.

We now consider the family of all simple submodules of M. By Proposition 1.7, there is a maximal collection $\mathscr{C}$ of simple submodules of M such that $\sum_{S \in \mathscr{C}} S$ is a direct sum. Put $K = \sum_{S \in \mathscr{C}} S$. If $K \neq M$, then there is a non-zero submodule L of M such that $M = K + L$ (d.s.). But then L contains a simple submodule $\bar{S}$ (say), and the sum $\bar{S} + \sum_{S \in \mathscr{C}} S$ is direct. This contradicts the maximality of $\mathscr{C}$. Hence $M = \sum_{S \in \mathscr{C}} S$ and (a) is established. $\square$

DEFINITION *A module M satisfying the equivalent conditions of Proposition 3.2 is said to be a 'semi-simple module'. A ring is* said to be '(*left*) *semi-simple*' if it is semi-simple as a left module over itself.

REMARK   There is a similar definition of a (right) semi-simple ring. The reason the words 'left' and 'right' are placed in parentheses is that the two concepts turn out to be equivalent. This is a consequence of the celebrated *Artin–Wedderburn theorem*. We shall assume this implicitly in that we shall refer simply to a 'semi-simple ring', although for our purposes a semi-simple ring R may be defined as one which can be written as the (direct) sum of a family of simple left R-ideals.

If a module is semi-simple, then so is every module which is isomorphic to it. A simple module is semi-simple, as is a zero module.

Considerable simplifications occur when we are dealing with semi-simple modules, as the next proposition shows.

PROPOSITION 3.3   *Let M be a semi-simple R-module. Then the following statements are equivalent:*

(a) *M is the direct sum of a finite family of simple submodules;*
(b) *M is Noetherian;*
(c) *M is Artinian;*
(d) *M is finitely generated.*

Proof. Proposition 1.18 Corollary says that (a) implies (b) and (c). Proposition 1.3 gives that (a) implies (d). Now suppose that (a) is false. Then M must be the direct sum of an infinite family of simple submodules. But then M has infinite strictly ascending and descending chains of submodules, so (b) and (c) are false.

It remains to show that (a) follows from (d). Suppose that $M = Rm_1 + Rm_2 + \ldots + Rm_n$, so that M is finitely generated. Since M is semi-simple, we can write

$$M = \sum_{\lambda \in \Lambda} S_\lambda \text{ (d.s.)},$$

where the S_λ are simple. We can now pick out a finite collection $\lambda_1, \lambda_2, \ldots, \lambda_r$ of elements of Λ such that

$$m_i \in S_{\lambda_1} + S_{\lambda_2} + \ldots + S_{\lambda_r}$$

for $1 \leqslant i \leqslant n$. But then

$$M = S_{\lambda_1} + S_{\lambda_2} + \ldots + S_{\lambda_r} \text{ (d.s.)}$$

and $\Lambda = \{\lambda_1, \lambda_2, \ldots, \lambda_r\}$. This gives (a). $\square$

COROLLARY *A semi-simple ring is the direct sum of a finite family of simple left ideals.*

Proof. This follows because a ring R is generated as an R-module by its identity element. $\square$

PROPOSITION 3.4 *Every submodule of a semi-simple module is semi-simple.*

Proof. This follows immediately from Lemma 3.1. $\square$

PROPOSITION 3.5 *A direct sum of semi-simple modules is semi-simple.*

Proof. This is immediate. $\square$

PROPOSITION 3.6 *A homomorphic image of a semi-simple module is semi-simple.*

Proof. Let M be a semi-simple R-module, let M' be an R-module and let $\phi: M \to M'$ be an epimorphism. Then $\operatorname{Ker}\phi$ is a direct summand of M, so that there is a submodule N of M such that $M = \operatorname{Ker}\phi + N$ (d.s.). Now $M' \approx M/\operatorname{Ker}\phi \approx N$ and N is semi-simple by Proposition 3.4. $\square$

The next result provides among other things a characterization of semi-simple rings in terms of injective modules.

PROPOSITION 3.7 *Let R be a ring. Then the following statements are equivalent:*

  (a) *R is semi-simple;*
  (b) *every R-module is semi-simple;*
  (c) *every R-module is injective;*
  (d) *every left ideal of R is injective.*

Proof. Since every R-module is a homomorphic image of a direct sum of copies of R (Proposition 1.4), (b) follows from (a) by means of Propositions 3.5 and 3.6.

Now assume (b) and consider an R-module M. Let M' be an extension of M. Then M' is semi-simple, so that M is a direct summand of M'. Theorem 2.15 now gives that M is injective, and (c) follows.

That (d) follows from (c) is clear, so we assume (d) and deduce (a). Let I be a left ideal of R. Then I is injective and so is a direct summand of R (Theorem 2.15). It follows that R is semi-simple. $\square$

In the expression of a semi-simple module as a direct sum of simple submodules, it may be asked what degree of uniqueness is obtained. In fact, if $\{S_i\}_{i \in I}$ and $\{T_j\}_{j \in J}$ are families of simple submodules of an R-module M such that

$$\sum_{i \in I} S_i \text{ (d.s.)} = \sum_{j \in J} T_j \text{ (d.s.)},$$

then there is a one–one correspondence between the two families such that corresponding simple modules are isomorphic. This is proved in the next section.

We shall actually establish a more general result. We showed in Proposition 1.23 that the ring of endomorphisms of a simple module is a division ring. Now the class of division rings is contained in the wider class of 'quasi-local' rings. We shall see that indecomposable injective modules as well as simple modules have quasi-local rings of endomorphisms. The uniqueness theorem that will be proved applies to direct sums of modules which have quasi-local rings of endomorphisms, and thus brings under one roof both the simple modules and the indecomposable injective modules.

3.2. Modules with quasi-local rings of endomorphisms

We first introduce the notion of a quasi-local ring.

LEMMA 3.8 *Let J be a subset of the ring R. Then the following statements are equivalent:*

(a) *J is the set of elements of R which do not have left inverses and is a left ideal of R;*

(b) *J is the set of elements of R which do not have right inverses and is a right ideal of R;*

(c) *J is the set of elements of R with neither a left nor a right inverse, the elements not in J are units and J is a two-sided ideal;*

(d) *J is the set of non-units, is non-empty, and is closed under addition.*

Note that a *unit* of R is an element with both a left and a right inverse. Let α be a unit of R. Then there are elements β, γ of R such that $\beta\alpha = \alpha\gamma = 1$. But then

$$\beta = \beta 1 = \beta(\alpha\gamma) = (\beta\alpha)\gamma = 1\gamma = \gamma.$$

Thus a left inverse of α and a right inverse of α are the same, so that α possesses a unique *inverse* α^{-1} such that

$$\alpha^{-1}\alpha = \alpha\alpha^{-1} = 1.$$

Before we prove Lemma 3.8, we shall make a definition and add some comments.

DEFINITION *A ring R which has a subset J satisfying the equivalent conditions (a), (b), (c) and (d) of Lemma 3.8 is said to be a 'quasi-local ring'.*

NOTES

(i) A quasi-local ring is necessarily non-trivial, since the only element of a trivial ring is a unit.

(ii) Condition (a) implies that J is the only maximal left ideal of R.

(iii) Condition (b) implies that J is the only maximal right ideal of R.

(iv) Condition (c) implies that J is the only maximal two-sided ideal of R.

(v) We deduce from (c) that a quasi-local ring has no elements with just one-sided inverses.

(vi) Condition (d) will be the most useful when we come to check that a given ring is quasi-local.

(vii) We have already noted that a division ring is quasi-local; in that case $J = 0$.

Proof of Lemma 3.8. We assume (a) and deduce (b). Consider an element x of J and suppose that x has a right inverse y, so that $xy = 1$. We shall derive a contradiction. Then

$$(1 - yx)y = y - yxy = y - y = 0.$$

Now $yx \in J$ because J is a left ideal, so that $1 - yx \notin J$, otherwise $1 \in J$. Hence $1 - yx$ has a left inverse z (say). Then

$$y = z(1 - yx)y = 0.$$

This provides the required contradiction and no element of J has a right inverse.

We shall now show that J is a right ideal of R. Let $r \in R$. Then Jr is a left ideal of R and $Jr \neq R$ because we have just shown that the elements of J do not have right inverses. Since J is the only maximal left ideal of R, it follows that $Jr \subseteq J$. This shows that J is a right ideal of R, since we already know that J, being a left ideal, is closed under addition.

Finally let s be an element of R not in J. We must show that s has a right inverse. Certainly s has a left inverse t (say), so that $ts = 1$. Also $t \notin J$, because J is a proper right ideal. Hence t has a left inverse u (say), so that $ut = 1$. Then $u = uts = s$, so that $st = 1$ and s has a right inverse t. This establishes (b). It follows by symmetry that (b) implies (a), so that (a) and (b) are equivalent. It is immediate that (a) and (b) together are equivalent to (c) and that (c) implies (d).

Before completing the proof of Lemma 3.8, we shall insert a subsidiary lemma.

LEMMA 3.9 *Let α, β be elements of a ring R and suppose that α and $\alpha\beta$ are units of R. Then β is also a unit.*

Proof. Let α and $\alpha\beta$ have respective inverses γ, δ, so that $\gamma\alpha = \alpha\gamma = 1$ and $\delta(\alpha\beta) = (\alpha\beta)\delta = 1$. Then

$$\beta(\delta\alpha) = \gamma\alpha\beta\delta\alpha = \gamma\alpha = 1.$$

Since $(\delta\alpha)\beta = 1$, this shows that β has inverse $\delta\alpha$. □

To complete the proof of Lemma 3.8, we assume (d) and deduce (a). Let $x \in J$ and $r \in R$. We shall show that $rx \in J$, which will show that J is a left ideal. If $r \notin J$, then $rx \in J$ by Lemma 3.9. Suppose that $r \in J$. Then $-r \in J$, so that $1 + r \notin J$ because J is closed under addition. It follows by Lemma 3.9 that $(1+r)x \in J$, and again $rx \in J$. No element of J can have a left inverse, because J is a left ideal and $1 \notin J$. On the other hand, the elements not in J certainly do have left inverses. We have now deduced (a) from (d). □

We shall want to consider the ring of endomorphisms of an injective module. The next lemma is useful in this context.

LEMMA 3.10 *Let E be an indecomposable injective R-module and let f belong to the ring of endomorphisms of E. Then f is a unit if and only if $\operatorname{Ker} f = 0$.*

Proof. The units of the ring of endomorphisms of E are just the R-isomorphisms from E to E. It is thus clear that, if f is a unit, then $\operatorname{Ker} f = 0$. Now suppose that $\operatorname{Ker} f = 0$. Then f is a monomorphism and $f(E)$ is an injective module. It follows from Theorem 2.15 that $f(E)$ is a direct summand of E, and $f(E) \neq 0$, so that $f(E) = E$. Hence f is also an epimorphism, and so is a unit of the ring of endomorphisms of E. $\square$

PROPOSITION 3.11 *Let E be an R-module such that its ring of endomorphisms is quasi-local. Then E is indecomposable.*

Proof. A zero module has as its ring of endomorphisms a trivial ring, so $E \neq 0$. Suppose that E is not indecomposable. Then there exist non-zero submodules E_1 and E_2 such that $E = E_1 + E_2$ (d.s.). Denote the respective projection and injection mappings by $\pi_1 \colon E \to E_1$, $\pi_2 \colon E \to E_2$, $\phi_1 \colon E_1 \to E$, $\phi_2 \colon E_2 \to E$. Then $\phi_1 \pi_1$ and $\phi_2 \pi_2$ are non-units of $\mathrm{Hom}_R(E, E)$, yet $\phi_1 \pi_1 + \phi_2 \pi_2 = \mathrm{id}_E$. It follows that the ring of endomorphisms of E is not quasi-local. $\square$

PROPOSITION 3.12 *Let E be an injective R-module. Then the following statements are equivalent:*

   (a) *E is indecomposable;*

   (b) *the ring of endomorphisms of E is quasi-local.*

Proof. We shall assume that E is indecomposable and deduce (b); because of Proposition 3.11, this will complete the proof. Since $E \neq 0$, its ring of endomorphisms is non-trivial. Let f and g be non-units of $\mathrm{Hom}_R(E, E)$. Then, by Lemma 3.10, $\mathrm{Ker} f \neq 0$ and $\mathrm{Ker} g \neq 0$, whence, by Proposition 2.28, $\mathrm{Ker} f \cap \mathrm{Ker} g \neq 0$. But $\mathrm{Ker} f \cap \mathrm{Ker} g \subseteq \mathrm{Ker}(f+g)$. It follows that $f+g$ is a non-unit. Hence the ring of endomorphisms of E is quasi-local. $\square$

We now come to the uniqueness theorem promised in the previous section. This is the celebrated *Krull–Schmidt–Remak–Azumaya theorem*. The proof uses none of the theory of injective modules developed so far.

THEOREM 3.13 *Let M be an R-module and suppose that $\{M_i\}_{i \in I}$ and $\{N_j\}_{j \in J}$ are families of submodules of M such that*

$$\sum_{i \in I} M_i \,(\text{d.s.}) = M = \sum_{j \in J} N_j \,(\text{d.s.}).$$

Suppose further that each M_i and each N_j has a quasi-local ring of endomorphisms. Then there is a one–one correspondence between the families $\{M_i\}_{i \in I}$ and $\{N_j\}_{j \in J}$ such that corresponding modules are isomorphic.

The lemma which follows is crucial for the proof of this theorem.

LEMMA 3.14 *Let M be an R-module and suppose that*

$$M = \sum_{i \in I} M_i \text{ (d.s.)} = N + N' \text{ (d.s.)},$$

where N and the M_i are submodules of M with quasi-local rings of endomorphisms. Then there exists $i \in I$ such that the combined mapping

$$M_i \xrightarrow{\text{inc}} M \xrightarrow{\text{proj}} N$$

is an isomorphism and

$$M = M_i + N' \text{ (d.s.)}.$$

Proof. Since N has a non-trivial ring of endomorphisms, $N \neq 0$ and there exists a non-zero element n in N. We can write

$$n = m_1 + m_2 + \ldots + m_s,$$

where, for $1 \leqslant k \leqslant s$, m_k is a non-zero element of M_{i_k}. We put

$$P = M_{i_1} + M_{i_2} + \ldots + M_{i_s}, \quad Q = \sum_{\substack{i \neq i_\alpha \\ 1 \leqslant \alpha \leqslant s}} M_i,$$

so that

$$M = P + Q \text{ (d.s.)}.$$

We now consider the combined mappings

$$\phi_1 \colon N \xrightarrow{\text{inc}} M \xrightarrow{\text{proj}} P \xrightarrow{\text{inc}} M \xrightarrow{\text{proj}} N,$$

$$\phi_2 \colon N \xrightarrow{\text{inc}} M \xrightarrow{\text{proj}} Q \xrightarrow{\text{inc}} M \xrightarrow{\text{proj}} N.$$

These are both endomorphisms of N and $\phi_1 + \phi_2 = \text{id}_N$. Since the ring of endomorphisms of N is quasi-local, it follows that one of ϕ_1, ϕ_2 must be an isomorphism. But it cannot be ϕ_2 because $\phi_2(n) = 0$, so ϕ_1 must be an isomorphism.

We now consider the endomorphisms of N given by

$$\psi_k \colon N \xrightarrow{\text{inc}} M \xrightarrow{\text{proj}} M_{i_k} \xrightarrow{\text{inc}} M \xrightarrow{\text{proj}} N$$

for $1 \leqslant k \leqslant s$. Now

$$\psi_1 + \psi_2 + \ldots + \psi_s = \phi_1,$$

which is an isomorphism, so that at least one of the ψ_k must be an isomorphism. This means that there exists $i \in I$ such that the combined mapping

$$\psi \colon N \xrightarrow{\text{inc}} M \xrightarrow{\text{proj}} M_i \xrightarrow{\text{inc}} M \xrightarrow{\text{proj}} N$$

is an isomorphism. We split this into two mappings, namely

$$\alpha\colon N \xrightarrow{\text{inc}} M \xrightarrow{\text{proj}} M_i \quad \text{and} \quad \beta\colon M_i \xrightarrow{\text{inc}} M \xrightarrow{\text{proj}} N,$$

so that $\psi = \beta\alpha$. This gives straightaway that β is an epimorphism. Consider $x \in M_i$. Then $\beta(x) \in N$, so that $\beta(x) = \beta\alpha(x')$ for some $x' \in N$, whence $x = \alpha(x') + (x - \alpha(x')) \in \alpha(N) + \operatorname{Ker}\beta$. Since

$$\alpha(N) \cap \operatorname{Ker}\beta = 0,$$

this gives that $\qquad M_i = \alpha(N) + \operatorname{Ker}\beta$ (d.s.).

By Proposition 3.11, M_i is indecomposable. Also, $\alpha(N) \neq 0$, so that $\operatorname{Ker}\beta = 0$ and β is an isomorphism. This establishes the first part of the lemma.

Now consider $y \in N$, and denote by $\pi\colon M \to N$ the projection mapping derived from the direct sum $M = N + N'$ (d.s.). Then there exists $y' \in M_i$ such that $\pi(y) = y = \beta(y') = \pi(y')$, so that $y - y' \in \operatorname{Ker}\pi = N'$. This gives that $y \in M_i + N'$ and shows that $M = M_i + N'$. But $M_i \cap N' = \operatorname{Ker}\beta = 0$, so the sum $M_i + N'$ is direct. Everything in the lemma is now proved. $\square$

In proving Theorem 3.13, we shall make use of two results from the theory of sets. One is the celebrated *Schroeder–Bernstein theorem*. This asserts that if X and Y are non-empty sets for which there are injections $X \to Y$ and $Y \to X$, then there is a bijection between X and Y. For the other, suppose we have an infinite family $\{X_i\}_{i \in I}$ of finite sets, not all empty. Then there is an injection

$$\bigcup_{i \in I} X_i \to I.$$

This is a result in the theory of cardinal numbers. We refer the reader to Bourbaki, *Elements of Mathematics, Theory of Sets* (Herman/Addison–Wesley, 1968) §6 no. 1 Theorem 2, Corollary 3 for details.

Proof of Theorem 3.13 Let the situation be as described in the statement of the theorem. We may suppose that $M \neq 0$, so that the index sets I and J are not empty. The index set I is partitioned as follows: the elements i, i' of I will belong to the same set of the partition if and only if $M_i \approx M_{i'}$. The same is done for the index set J. It follows from Lemma 3.14 that there is a one–one correspondence between the sets of the two partitions; the sets I_0 and

J_0 correspond if $M_i \approx N_j$ wherever $i \in I_0$ and $j \in J_0$. Consider corresponding sets I_0 and J_0. The theorem will follow if we can show that there is a bijection between I_0 and J_0.

Consider an element j_1 of J_0. By Lemma 3.14, there exists $i_1 \in I_0$ such that

$$M = M_{i_1} + (\sum_{j \neq j_1} N_j) \quad (\text{d.s.}).$$

Now consider an element j_2 of J_0, $j_2 \neq j_1$. There exists $i_2 \in I_0$ such that

$$M = M_{i_1} + M_{i_2} + (\sum_{j \neq j_1, j_2} N_j) \quad (\text{d.s.})$$

Note that $i_2 \neq i_1$. Continuing in this way, we see that, for any finite set of distinct elements of J_0, there is a finite set of an equal number of distinct elements of I_0. Of course, the roles of I_0 and J_0 can be interchanged, and this shows that, if one of I_0 and J_0 is finite, then so is the other and they have the same number of elements.

It remains for us to consider the case when I_0 and J_0 are both infinite. Consider $i \in I_0$. For each $j \in J$, we have a mapping

$$\theta_j: M_i \xrightarrow{\text{inc}} M \xrightarrow{\text{proj}} N_j.$$

We denote by $J(i)$ the set of all $j \in J$ such that the mapping θ_j is an isomorphism. Then $J(i) \subseteq J_0$. Let m be a non-zero element of M_i. Then we can write

$$m = n_1 + n_2 + \ldots + n_t,$$

where $n_k \in N_{j_k}$ $(1 \leqslant k \leqslant t)$ and each n_k is non-zero. Then $m \in \mathrm{Ker}\, \theta_j$ whenever $j \neq j_1, j_2, \ldots, j_t$. It follows that $J(i)$ is a finite set. Moreover, by Lemma 3.14, each element of J_0 belongs to $J(i)$ for some $i \in I_0$. It follows that

$$\bigcup_{i \in I_0} J(i) = J_0.$$

The set-theoretic result that we quoted prior to the present proof now enables us to deduce that there is an injection $J_0 \to I_0$. By symmetry, there is also an injection $I_0 \to J_0$. The Schroeder–Bernstein theorem now gives a bijection between I_0 and J_0. $\square$

We set down in a corollary the cases of Theorem 3.13 that especially interest us.

Corollary   *Let $\{E_i\}_{i \in I}$ and $\{F_j\}_{j \in J}$ be families of R-modules such that*

$$\bigoplus_{i \in I} E_i \approx \bigoplus_{j \in J} F_j.$$

Suppose that either (a) the E_i and the F_j are simple, or (b) the E_i and the F_j are indecomposable injective modules. Then there is a one–one correspondence between the two families such that corresponding modules are isomorphic.

Proof. This follows from Theorem 3.13 by means of Proposition 1.23 in case (a) and Proposition 3.12 in case (b). □

3.3   The socle of a module

Definition   *Let M be an R-module. We denote by $S(M)$ the sum of all the simple submodules of M and call $S(M)$ the 'socle' of M. In the case when M has no simple submodules, then, with our usual convention, $S(M)$ becomes the zero submodule of M.*

The socle of a module is sem-simple. In fact, it is the unique maximal semi-simple submodule of the module. Also, an isomorphism between modules restricts to an isomorphism between their socles.

Proposition 3.15   *Let $\{E_i\}_{i \in I}$ be a family of R-modules. Then*

$$S\left(\bigoplus_{i \in I} E_i\right) = \bigoplus_{i \in I} S(E_i).$$

Proof. The submodule $\bigoplus_{i \in I} S(E_i)$ of $\bigoplus_{i \in I} E_i$ is semi-simple, so that

$$\bigoplus_{i \in I} S(E_i) \subseteq S\left(\bigoplus_{i \in I} E_i\right).$$

On the other hand, consider a simple submodule of $\bigoplus_{i \in I} E_i$. This is singly generated, say by x. Suppose that x has a non-zero component x_j in E_j. There is an epimorphism $Rx \to Rx_j$ which maps rx to rx_j $(r \in R)$, so that Rx_j is a simple module. Hence $Rx_j \subseteq S(E_j)$. It follows that

$$Rx \subseteq \bigoplus_{i \in I} S(E_i),$$

which shows that

$$S\left(\bigoplus_{i \in I} E_i\right) \subseteq \bigoplus_{i \in I} S(E_i). \quad \Box$$

PROPOSITION 3.16   *Let M' be a submodule of the R-module M. Then $S(M')$ is a submodule of $S(M)$. Moreover, $S(M') = S(M)$ if M is an essential extension of M'. In particular, $S(E(M)) = S(M)$.*

Proof. The first assertion is clear. Now let M be an essential extension of M', and consider a simple submodule S of M. Then $S \cap M' \neq 0$, so that $S \cap M' = S$, i.e. $S \subseteq M'$. It follows from this that $S(M') = S(M)$, as required. □

PROPOSITION 3.17   *Let M be an R-module. Then the following statements are equivalent:*

(a) *every non-zero submodule of M contains a simple submodule;*
(b) *M is an essential extension of $S(M)$.*
Moreover, when (a) and (b) hold, then $E(M) = E(S(M))$.

Proof. Assume (a). If M' is a non-zero submodule of M, then M' contains a simple submodule S (say), so that $M' \cap S(M) \neq 0$. It follows that M is an essential extension of $S(M)$.

Now assume (b). If M' is a non-zero submodule of M, then $M' \cap S(M) \neq 0$. Then $M' \cap S(M)$, being a submodule of the semi-simple module $S(M)$, is also semi-simple (Proposition 3.4), so $M' \cap S(M)$ has a simple submodule. It follows that M' also has a simple submodule.

The final remark follows from Proposition 2.22. □

COROLLARY   *Let M be an R-module. If either* (i) *R is an Artinian ring, or* (ii) *M is an Artinian module, then M is an essential extension of $S(M)$ and $E(M) = E(S(M))$.*

Proof. This follows by Propositions 1.20 and 1.21. □

3.4   Finitely embedded modules

A Noetherian module may be characterized as one such that every submodule is finitely generated. When this was discussed in Section 1.8, we were not able to provide a dual description of Artinian modules. This will be done in the present section.

DEFINITION   *An R-module M is said to be 'finitely embedded' if there exist finitely many simple modules $S_1, S_2, \ldots, S_k$ such that*

$$E(M) \approx E(S_1) \oplus E(S_2) \oplus \ldots \oplus E(S_k).$$

By convention, a zero module is finitely embedded, for in that case we take no simple modules. Note that the property of being finitely embedded is preserved under isomorphism.

PROPOSITION 3.18   *An R-module M is finitely embedded if and only if the following conditions hold:*
(a) *M is an essential extension of S(M);*
(b) *S(M) is finitely generated.*
Proof. Suppose first that M is finitely embedded, and write

$$E(M) \approx E(S_1) \oplus E(S_2) \oplus ... \oplus E(S_k), \qquad (3.4.1)$$

where $S_1, S_2, ..., S_k$ are simple modules. This isomorphism restricts to an isomorphism between the socles of the respective modules. By Proposition 3.16, $E(M)$ has socle $S(M)$; by Proposition 3.15,

$$S(E(S_1) \oplus ... \oplus E(S_k)) = S(E(S_1)) \oplus ... \oplus S(E(S_k))$$
$$= S_1 \oplus ... \oplus S_k.$$

Hence the isomorphism (3.4.1) restricts to an isomorphism

$$S(M) \approx S_1 \oplus ... \oplus S_k.$$

This shows that $S(M)$ is finitely generated. Further, Proposition 2.16 shows that $E(S_1) \oplus ... \oplus E(S_k)$ is an essential extension of $S_1 \oplus ... \oplus S_k$, whence $E(M)$ is an essential extension of $S(M)$. It follows that M is an essential extension of $S(M)$. This establishes (a) and (b).

Now suppose that (a) and (b) hold. By Proposition 3.3, we can write
$$S(M) = S_1 + ... + S_k \text{ (d.s.)},$$

where $S_1, ..., S_k$ are simple submodules of M. Hence, by Propositions 3.17 and 2.23,

$$E(M) = E(S(M)) \approx E(S_1) \oplus ... \oplus E(S_k).$$

Thus M is finitely embedded. $\square$

The next two pairs of results bring out the duality between the notions of finitely generated and finitely embedded modules. We first give a pair of definitions.

DEFINITION *Let M be an R-module. A family $\{M_i\}_{i \in I}$ of submodules of M is said to be a 'direct system' if, for any finite number of elements $i_1, \ldots, i_k$ of I, there is an element i_0 in I such that*

$$M_{i_0} \supseteq M_{i_1} + \ldots + M_{i_k}.$$

DEFINITION *Let M be an R-module. A family $\{M_i\}_{i \in I}$ of submodules of M is said to be an 'inverse system' if, for any finite number of elements $i_1, \ldots, i_k$ of I, there is an element i_0 in I such that*

$$M_{i_0} \subseteq M_{i_1} \cap \ldots \cap M_{i_k}.$$

Note that direct and inverse systems of submodules are non-empty.

PROPOSITION 3.19' *Let M be an R-module. Then the following statements are equivalent:*

(*a*) *M is finitely generated;*

(*b*) *every direct system of proper submodules of M is bounded above by a proper submodule of M.*

PROPOSITION 3.19 *Let M be an R-module. Then the following statements are equivalent:*

(*a*) *M is finitely embedded;*

(*b*) *every inverse system of non-zero submodules of M is bounded below by a non-zero submodule of M.*

REMARK To say that a family of submodules of M is bounded above (resp. below) by the submodule M' of M is to say that M' contains (resp. is contained in) every submodule of the family. Although we are mainly interested in finitely embedded modules here, we shall give proofs of both results.

Proof of Proposition 3.19'. Suppose first that M is finitely generated, say by $m_1, m_2, \ldots, m_n$, yet that M has a direct system $\{M_i\}_{i \in I}$ of proper submodules which is not bounded above by a proper submodule. Then $\sum\limits_{i \in I} M_i = M$. Now each m_k is a finite sum of elements from the M_i, and there are only finitely many m_k, so that there is a finite subset I' of I such that $\sum\limits_{i \in I'} M_i = M$. But $\{M_i\}_{i \in I}$ is a direct system, so there is an element $i_0 \in I$ such that $\sum\limits_{i \in I'} M_i \subseteq M_{i_0}$. But M_{i_0} is proper. This gives a contradiction and shows that (*a*) implies (*b*).

Now assume (*b*), but suppose that M is not finitely generated. Consider the collection of all finitely generated submodules of M. This is a direct system of proper submodules of M and so has a proper submodule M' as an upper bound. This means that there is an element $m \in M$, $m \notin M'$. But the submodule Rm of M is finitely generated, and so should be contained in M'. This gives a contradiction and shows that (*b*) implies (*a*). $\square$

Proof of Proposition 3.19. Suppose first that M is finitely embedded, and consider an inverse system $\{M_i\}_{i \in I}$ of non-zero submodules of M. By Proposition 3.18, M is an essential extension of $S(M)$, so $M_i \cap S(M) \neq 0$ for each $i \in I$. Also, $S(M)$ is finitely generated and so Artinian (Proposition 3.3). Hence there is a minimal member $M_{i_0} \cap S(M)$ among the $M_i \cap S(M)$. Consider any $i \in I$. Since $\{M_i\}_{i \in I}$ is an inverse system, there exists $i' \in I$ such that $M_{i'} \subseteq M_i \cap M_{i_0}$. Then $M_{i'} \cap S(M) \subseteq M_{i_0} \cap S(M)$, so that $M_{i'} \cap S(M) = M_{i_0} \cap S(M)$. But then $M_{i_0} \cap S(M) \subseteq M_{i'} \subseteq M_i$ so that $M_{i_0} \cap S(M)$ is a lower bound for the inverse system. Thus (*a*) implies (*b*).

Conversely, we assume that every inverse system of non-zero submodules of M is bounded below by a non-zero submodule. We again employ the criterion for finitely embedded modules given in Proposition 3.18. We first show that M is an essential extension of $S(M)$. Consider a non-zero submodule N of M. Denote by Ω the collection of all non-zero submodules of N, partially ordered by the opposite of inclusion. Consider a totally ordered subset Ω' of Ω. Then Ω' is an inverse system and so is bounded above (remember that the ordering is the opposite of inclusion). It follows by Zorn's Lemma that Ω has a maximal member, i.e. N has a simple submodule S (say). But now $N \cap S(M) \supseteq S \neq 0$, and M is an essential extension of $S(M)$.

Finally, suppose that $S(M)$ is the direct sum of an infinite family of simple submodules, and let $\{S_i\}_{i=1}^{\infty}$ be a countable subfamily of this family. Then the family of non-zero submodules $\left\{ \sum\limits_{i=n}^{\infty} S_i \right\}_{n=1}^{\infty}$ is an inverse system. But the intersection of all the submodules of this system is zero, so it cannot be bounded below by a non-zero submodule. This gives a contradiction and shows that $S(M)$ is the direct sum of a finite number of simple sub-modules, i.e. $S(M)$ is finitely generated.□

For the next pair of results, we have an exact sequence

$$0 \to A' \xrightarrow{\phi} A \xrightarrow{\psi} A'' \to 0$$

of R-modules.

PROPOSITION 3.20' *If A is finitely generated, then so is A''. If A' and A'' are finitely generated, then so is A.*

PROPOSITION 3.20 *If A is finitely embedded, then so is A'. In particular, every submodule of a finitely embedded module is finitely embedded. Also, if A' and A'' are finitely embedded, then so is A.*

Proof of Proposition 3.20'. If A is generated by the elements $a_1, a_2, ..., a_n$, then A'' is generated by $\psi(a_1), \psi(a_2), ..., \psi(a_n)$.

Now suppose that

$$A' = Ra_1' + Ra_2' + ... + Ra_s' \quad \text{and} \quad A'' = Ra_1'' + Ra_2'' + ... + Ra_t''.$$

Since ψ is an epimorphism, there exist $a_1, a_2, ..., a_t \in A$ such that $a_i'' = \psi(a_i)$ for $1 \leqslant i \leqslant t$. Consider $a \in A$. We can write

$$\psi(a) = \sum_{i=1}^{t} r_i \psi(a_i)$$

where the $r_i \in R$, so that

$$a - \sum_{i=1}^{t} r_i a_i \in \operatorname{Ker} \psi = \operatorname{Im} \phi.$$

Thus we can write

$$a - \sum_{i=1}^{t} r_i a_i = \phi \left(\sum_{j=1}^{s} r_j' a_j' \right),$$

where the $r_j' \in R$. This shows that

$$A = Ra_1 + ... + Ra_t + R\phi(a_1') + ... + R\phi(a_s')$$

and so is finitely generated.□

Proof of Proposition 3.20. It follows from Proposition 3.19 that, if A is finitely embedded, then so is A'. Conversely, suppose that A' and A'' are finitely embedded, so that we can write

$$E(A') \approx E(S_1') \oplus ... \oplus E(S_k'),$$

$$E(A'') \approx E(S_1'') \oplus ... \oplus E(S_l''),$$

where the S_i' and S_j'' are simple modules. Then

$$E(E(A') \oplus E(A'')) = E(A') \oplus E(A'')$$

$$\approx E(S_1') \oplus \dots \oplus E(S_k') \oplus E(S_1'') \oplus \dots \oplus E(S_l''),$$

which shows that $E(A') \oplus E(A'')$ is also finitely embedded. Since $E(A')$ is injective, we can complete the diagram Fig. 3.1 by a homomorphism ϕ' as shown. We now define the mapping

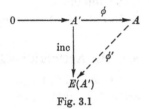

Fig. 3.1

$$\theta \colon A \to E(A') \oplus E(A'')$$

by $\theta(a) = (\phi'(a), \psi(a))$, where $a \in A$. Then θ is an R-homomorphism. But suppose that $\theta(a) = 0$ for some $a \in A$. Then $\psi(a) = 0$, so there exists $a' \in A'$ such that $a = \phi(a')$. Then

$$a' = \phi'\phi(a') = \phi'(a) = 0,$$

so that $a = 0$, i.e. θ is a monomorphism. It now follows from the first part that A is finitely embedded.$\square$

The next result provides the missing characterization of Artinian modules.

THEOREM 3.21 *Let M be an R-module. Then the following statements are equivalent:*
(a) *M is Artinian;*
(b) *every factor module of M is finitely embedded.*

Proof. Suppose first that M is Artinian, and let M' be a factor module of M. Then M' is also Artinian (Proposition 1.17), and Proposition 3.17 Corollary shows that M' is an essential extension of $S(M')$. Also, $S(M')$ is Artinian and so finitely generated (Proposition 3.3). It follows by Proposition 3.18 that M' is finitely embedded.

Conversely, suppose that every factor module of M is finitely embedded, and consider a descending chain

$$A_1 \supseteq A_2 \supseteq A_3 \supseteq \dots \tag{3.4.2}$$

of submodules of M. Put $A = \bigcap\limits_{n=1}^{\infty} A_n$, so that A is also a submodule of M. If the chain (3.4.2) does not terminate, then the module M/A has an inverse system

$$\{A_1/A, A_2/A, A_3/A, ...\}$$

of non-zero submodules which is not bounded below by a non-zero submodule. This contradicts Proposition 3.19, so the chain (3.4.2) must indeed terminate and M is Artinian. $\square$

3.5   An Artinian ring is Noetherian

For modules, the ascending and descending chain conditions are independent of each other. This is not the case if we restrict ourselves to rings however. We shall use the results of the previous section to show that, for a ring, the descending chain condition implies the ascending chain condition, i.e. an Artinian ring is Noetherian. Recall that, by an 'Artinian' (resp. 'Noetherian') ring, what we really mean is a *left* Artinian (resp. Noetherian) ring.

We denote by J the set of all elements r of R such that $rS = 0$ for every simple R-module S. Then J is a *two-sided* ideal of R and is called the *Jacobson radical* of R. Further, J *is the intersection of all the maximal left ideals of R.* For let $r \in J$. Then, for every maximal left ideal M of R, $r(R/M) = 0$, whence $r \in M$. Conversely, let r be an element of every maximal left ideal of R, and consider an arbitrary simple R-module S. Consider an arbitrary non-zero element s of S. Then $S = Rs \approx R/(0:s)$. Now $0:s$ is a maximal left ideal of R, so that $r \in 0:s$, whence $rs = 0$. It follows that $rS = 0$. This in turn means that $r \in J$.

In fact, J is also the intersection of all the maximal right ideals of R. Since we shall not be using this last fact, we leave its proof to the reader. (See DGN *Lessons* Chapter 7, Theorem 3.)

Because J is a two-sided ideal of R, we can form the residue class ring R/J. This is also a left R-module. We recall that the left ideals of R/J and its (left) submodules are one and the same. Hence its simple left ideals and its simple submodules are also one and the same.

PROPOSITION 3.22   *Let R be an Artinian ring. Then the Jacobson radical J of R is the intersection of a finite number of maximal left ideals of R.*

Proof. Theorem 3.21 shows that the R-module R/J is finitely embedded. Consider the collection of all finite intersections of submodules of R/J of the form M/J, where M is a maximal left ideal of R. This is an inverse system which is not bounded below by a non-zero R-module, since the intersection of all the M/J is zero. Because of Proposition 3.19, the only possibility is that one of the members of the collection is zero, i.e. there exist maximal left ideals $M_1, M_2, ..., M_n$ such that

$$(M_1/J) \cap (M_2/J) \cap ... \cap (M_n/J) = 0.$$

Thus $\qquad\qquad J = M_1 \cap M_2 \cap ... \cap M_n,$

as required.□

PROPOSITION 3.23 *Let R be an Artinian ring and let J be the Jacobson radical of R. Then R/J is a semi-simple ring.*

Proof. By Proposition 3.22, there are maximal left ideals $M_1, ..., M_n$ of R such that

$$J = M_1 \cap ... \cap M_n.$$

We define a mapping

$$\phi\colon R/J \to (R/M_1) \oplus ... \oplus (R/M_n)$$

by $\qquad\qquad \phi(r+J) = (r+M_1, ..., r+M_n) \quad (r \in R).$

This is a monomorphism of R-modules. Now

$$(R/M_1) \oplus ... \oplus (R/M_n)$$

is a semi-simple R-module. Thus R/J is a semi-simple R-module (Proposition 3.4) and so also a semi-simple ring.□

We recall that, if n is a positive integer, then J^n denotes the set of all finite sums of elements of R of the form $r_1 r_2 ... r_n$, where each r_i belongs to J. Then J^n is a two-sided ideal of R. We also put $J^0 = R$.

PROPOSITION 3.24 *Let R be an Artinian ring and let J be the Jacobson radical of R. Then there is a positive integer k such that $J^k = 0$.*

Proof. Since R is Artinian, the collection of all positive powers of J has a minimal member, which we denote by $I = J^k$. We note that I is a two-sided ideal of R and that $I^2 = J^{2k} = J^k = I$. Suppose that $I \neq 0$ and put

$$I' = \{r \in R \colon Ir = 0\}.$$

Then I' is a proper two-sided ideal of R. By Proposition 1.22, R/I' is an Artinian ring and so has a simple left ideal. Suppose that the element ξ of R/I' generates such an ideal. Then ξ also generates a simple R-module, so that $J\xi = 0$, by the very definition of J. Let $x \in R$ be a representative of the coset ξ. Then $Ix \subseteq Jx \subseteq I'$, so that $I^2x = 0$. But $I^2 = I$, so $Ix = 0$, whence $x \in I'$ and $\xi = 0$. This is just not so, because ξ generates a simple R-module. It follows that $J^k = I = 0.\ \square$

THEOREM 3.25   *Let R be an Artinian ring and M a left R-module. Then the following statements are equivalent:*
(a) *M is Noetherian;*
(b) *M is Artinian.*

Proof. Let J be the Jacobson radical of R. By Proposition 3.24, there is a positive integer k such that $J^k = 0$. Consider the sequence of submodules

$$M = J^0M \supseteq JM \supseteq J^2M \supseteq \ldots \supseteq J^kM = 0 \qquad (3.5.1)$$

of M. The factor module $J^{i-1}M/J^iM$, for $1 \leqslant i \leqslant k$, is annihilated by J, so it has the structure of an (R/J)-module. By Proposition 3.23, R/J is a semi-simple ring, so that $J^{i-1}M/J^iM$ is a semi-simple (R/J)-module (Proposition 3.7).

Suppose that M is Noetherian (resp. Artinian). Then

$$J^{i-1}M/J^iM$$

is Noetherian (resp. Artinian) as an R-module and so also as an (R/J)-module. It follows by Proposition 3.3 that $J^{i-1}M/J^iM$ is Artinian (resp. Noetherian), first as an (R/J)-module but then as an R-module. It follows from (3.5.1) that M is Artinian (resp. Noetherian).$\square$

If we put $M = R$, we obtain the result:

COROLLARY   *Every (left) Artinian ring is (left) Noetherian.*$\square$

Exercises on Chapter 3

3.1   An R-module is said to be *uniserial* if it has only finitely many submodules and these are totally ordered. Show that the ring of endomorphisms of a uniserial module $A \neq 0$ is quasi-local.

3.2 Let R be a left Ore domain (see Exercise 2.9) and let E be an injective envelope of R when R is considered as a left module over itself. Show that

(i) if two endomorphisms of E agree on a non-zero element of R, then they are equal;

(ii) the ring Q of endomorphisms of E is a division ring;

(iii) the opposite ring R^0 of R is isomorphic to the subring S of Q consisting of all endomorphisms f of E such that $f(R) \subseteq R$;

(iv) every element of Q can be written in the form fg^{-1}, where $f, g \in S$ and $g \neq 0$.

(The ring Q^0 is called a *left quotient ring* of R.)

Note. The *opposite ring* R^0 of R has the same elements and the same addition as R, but its multiplication o is defined by $a \circ b = ba$, where $a, b \in R$.

3.3 Show that a domain R has a left quotient ring if and only if it is a left Ore domain (see Exercise 3.2).

3.4 Let M be an R-module and let $\{M_i\}_{i \in I}$ and $\{N_j\}_{j \in J}$ be families of submodules of M whose respective sums are direct. Suppose further that each M_i and each N_j has a quasi-local ring of endomorphisms and that, for each finite subset I' resp. J' of I resp. J, $\sum\limits_{i \in I'} M_i$ and $\sum\limits_{j \in J'} N_j$ are direct summands of M. Finally, suppose that M is an essential extension of both $\sum\limits_{i \in I} M_i$ and $\sum\limits_{j \in J} N_j$. Prove that there is a bijection $\phi : I \to J$ such that, for every $i \in I$, $M_i \approx N_{\phi(i)}$. [*Hint*: Adapt the proof of Theorem 3.13.]

3.5 Let $\{E_i\}_{i \in I}$ and $\{F_j\}_{j \in J}$ be families of indecomposable injective R-modules such that

$$E(\bigoplus_{i \in I} E_i) \approx E(\bigoplus_{j \in J} F_j).$$

Show that there is a bijection $\phi : I \to J$ such that, for each $i \in I$, $E_i \approx F_{\phi(i)}$. [*Hint:* Use Exercise 3.4.]

3.6 Show that an R-module M is finitely embedded if and only if it contains an Artinian submodule N such that M is an essential extension of N.

3.7 Use Proposition 3.18 to show that every submodule of a finitely embedded R-module is finitely embedded. (See Proposition 3.20.)

3.8 Show that every essential extension of a finitely embedded module is finitely embedded.

3.9 Show that the ring R is left Artinian if and only if every finitely generated left R-module is finitely embedded.

3.10 Let R be a commutative ring, A a finitely generated R-module, and E a finitely embedded R-module. Show that the R-module $\mathrm{Hom}_R(A, E)$ is finitely embedded.

3.11 Let R be a ring and let $\alpha \in R$. Establish the equivalence of the following statements:

(i) α annihilates every simple left R-module (i.e. for every simple left R-module S, $\alpha S = 0$);

(ii) α annihilates every simple right R-module (i.e. for every simple right R-module T, $T\alpha = 0$);

(iii) α belongs to every maximal left ideal of R;

(iv) α belongs to every maximal right ideal of R;

(v) $1 - r\alpha$ has a left inverse for all $r \in R$;

(vi) $1 - \alpha r$ has a right inverse for all $r \in R$;

(vii) $1 - r\alpha r'$ has a two-sided inverse for all $r, r' \in R$.

Deduce that the Jacobson radical of R is also the intersection of all the maximal right ideals of R.

3.12 Let E be an injective R-module and let S be the ring of endomorphisms of E. Let I be the set of all endomorphisms f of E such that E is an injective envelope of $\mathrm{Ker}\, f$. Show that

(i) I is a two-sided ideal of S;

(ii) an element f of S belongs to I if and only if $\mathrm{id}_E - gf$ is a unit of S for all $g \in S$ (so that I is the Jacobson radical of S – see Exercise 3.11);

(iii) S/I is a regular ring (see Exercise 2.14).

Notes on Chapter 3

In view of the results in this chapter about the duality that exists between finitely generated and finitely embedded modules, it may be thought that every result about finitely generated modules dualizes to finitely embedded modules. This is not so. For example, every module is a direct limit of finitely generated modules, but not every module is an inverse limit of finitely embedded modules. The reason for this lack of complete duality will be clear from a look at the category of modules; it satisfies A. Grothendieck's axiom $AB5$, namely

$$A \cap (\bigcup_{i \in I} B_i) = \bigcup_{i \in I} (A \cap B_i),$$

where $\{B_i\}_{i \in I}$ is a totally ordered family of submodules of a module M and A is a submodule of M, and this property of modules is not dualizable.

4. Injective modules and chain conditions

4.1 Characterizations of Noetherian and Artinian rings

In Proposition 2.3, we showed that, if $\{E_i\}_{i \in I}$ is a family of R-modules whose direct sum is injective, then each E_i must be injective. But we cannot in general say that, if each E_i is injective, then $\underset{i \in I}{\oplus} E_i$ is also injective; although this does follow if the index set I is finite. We shall show here that it follows for an arbitrary family if and only if the ring R is Noetherian, so that this provides a characterization of Noetherian rings. We recall that a 'Noetherian' (resp. 'Artinian') ring is to mean a *left* Noetherian (resp. *left* Artinian) ring. Throughout the present chapter, the symbol R will continue to denote a ring. We recall also that, if A is an R-module, then $E(A)$ denotes an injective envelope of A.

THEOREM 4.1 *The following statements are equivalent:*

(a) R is a Noetherian ring;

(b) every direct sum of injective R-modules is injective;

(c) every direct sum of a countably infinite family of injective envelopes of simple R-modules is injective.

Proof. Assume first that R is a Noetherian ring, let $\{E_i\}_{i \in I}$ be a family of injective R-modules, and put $E = \underset{i \in I}{\oplus} E_i$. Consider the diagram Fig. 4.1 where B is a left ideal of R. Now B is finitely generated because R is Noetherian. It follows that there exists a finite subset J of I such that $f(B) \subseteq E' = \underset{j \in J}{\oplus} E_j$. We thus have Fig. 4.2. But E' is injective so there exists an R-homomorphism $\phi': R \to E'$ such that the resulting diagram shown is commutative. The original diagram is now completed by $\phi: R \to E$ which is ϕ' followed by the inclusion mapping. Hence E is injective. Thus (a) implies (b). It is immediate that (b) implies (c).

Now assume (c) and also the existence of an infinite strictly increasing sequence

$$A_1 \subset A_2 \subset \dots \subset A_n \subset \dots$$

of left ideals of R. We shall derive a contradiction. Put

$$A = \bigcup_{n=1}^{\infty} A_n.$$

Then A is a left ideal of R and, for each k, $A/A_k \neq 0$. Proposition 2.24 shows that there exists a simple module S_k and a non-zero

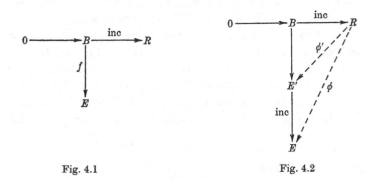

Fig. 4.1   Fig. 4.2

homomorphism $\alpha_k: A/A_k \to E(S_k)$. We denote by $\phi_k: A \to E(S_k)$ the mapping obtained by composing the natural mapping $A \to A/A_k$ with α_k. Then ϕ_k is non-zero. We now define the mapping

$$\phi: A \to \bigoplus_{k=1}^{\infty} E(S_k) = N \quad \text{(say)}$$

by
$$\phi(r) = \{\phi_k(r)\}_{k=1}^{\infty} \quad (r \in A).$$

Note that, for any $r \in A$, there exists k_0 such that $r \in A_k$ for all $k \geqslant k_0$, so that $\phi_k(r) = 0$ for all $k \geqslant k_0$ and ϕ is a well-defined R-homomorphism. By hypothesis, N is injective, so ϕ can be extended to a homomorphism $\psi: R \to N$. But R is singly generated as an R-module, so there exists an integer n such that $\psi(R) \subseteq \bigoplus_{k=1}^{n} E(S_k)$. This means that ϕ_k is the zero mapping for all $k > n$, which is a contradiction. Thus (c) implies (a). $\square$

The statement of Proposition 2.23 can now be strengthened when the ring is Noetherian, in that it holds for arbitrary families of modules, rather than just for finite families.

PROPOSITION 4.2   *Let R be a Noetherian ring and let $\{A_i\}_{i \in I}$ be a family of R-modules. Then $\bigoplus_{i \in I} E(A_i)$ is an injective envelope of $\bigoplus_{i \in I} A_i$, i.e.*

$$E(\bigoplus_{i \in I} A_i) = \bigoplus_{i \in I} E(A_i).$$

Proof. The proof is similar to that of Proposition 2.23. □

LEMMA 4.3   *Let M be a non-zero Noetherian R-module. Then M has a submodule K such that $E(K)$ is indecomposable.*

Proof. Suppose the contrary. Then, in particular, $E(M)$ is not indecomposable, so there exist non-zero submodules B_1, C_1 of M such that $B_1 \cap C_1 = 0$. But $E(B_1)$ is not indecomposable, so there exist non-zero submodules B_2, C_2 of B_1 such that $B_2 \cap C_2 = 0$. Continue in this way. At the nth stage ($n > 1$), we have non-zero submodules B_n, C_n of B_{n-1} such that $B_n \cap C_n = 0$. Suppose that

$$C_1 + C_2 + \ldots + C_{n-1} = C_1 + C_2 + \ldots + C_n$$

for some $n > 1$. Then

$$C_n \subseteq C_1 + C_2 + \ldots + C_{n-1}.$$

Consider an arbitrary element $x \in C_n$. There exist

$$x_1 \in C_1, \ldots, x_{n-1} \in C_{n-1}$$

such that $x_1 + \ldots + x_{n-1} + x = 0.$

Now $x_1 \in C_1 \cap (C_2 + \ldots + C_n) \subseteq C_1 \cap B_1 = 0,$

so $x_1 = 0$. Then

$$x_2 \in C_2 \cap (C_3 + \ldots + C_n) \subseteq C_2 \cap B_2 = 0,$$

so $x_2 = 0$. Continuing in this way, we see that $x = 0$. This contradicts the fact that $C_n \neq 0$. Thus we have a strictly ascending chain

$$C_1 \subset C_1 + C_2 \subset C_1 + C_2 + C_3 \subset \ldots$$

of submodules of M. This contradicts the fact that M is Noetherian. □

COROLLARY   *Let R be a Noetherian ring and let M be a non-zero R-module. Then M has a submodule K such that $E(K)$ is indecomposable.*

Proof. Consider any non-zero finitely generated submodule of M. By Proposition 1.19 this will be Noetherian and so has a submodule K such that $E(K)$ is indecomposable. □

Before we state the next result, we must introduce a notion which depends on the theory of cardinal numbers. We denote the cardinal of a set X by $|X|$. Let M be an R-module, and consider its socle, $S(M)$, which may be written as

$$S(M) = \sum_{i \in I} S_i \quad \text{(d.s.)},$$

where the S_i are simple submodules. By Theorem 3.13 Corollary, the cardinal number $|I|$ depends only on the module M. We shall denote this cardinal number by $c(M)$. Note that $c(M) = 0$ if and only if $S(M) = 0$. Also, if M is an indecomposable injective R-module, then its zero submodule is irreducible and $c(M)$ is either 0 or 1.

THEOREM 4.4   *The following statements are equivalent:*

(a) R is a Noetherian ring;

(b) every injective R-module is a direct sum of indecomposable (injective) R-modules;

(c) there is a cardinal number κ such that every injective R-module is of the form $\sum_{i \in I} M_i$ (d.s.), where $c(M_i) \leqslant \kappa$ for all $i \in I$.

Proof. Assume that R is a Noetherian ring and let E be an injective R-module. By Proposition 1.7, E has a family $\{E_i\}_{i \in I}$ of indecomposable injective submodules which is maximal with respect to the property that its sum is direct. By Theorem 4.1, $\sum_{i \in I} E_i$ is injective, and so is a direct summand of E. We write $E = (\sum_{i \in I} E_i) + E'$ (d.s.), where E' is a submodule of E. Then E' is also injective. If $E' \neq 0$, it follows from Lemma 4.3 Corollary that E' has a submodule K such that $E(K)$ is indecomposable. By Proposition 2.19, $E(K)$ may be chosen to be a submodule of E'. But in this case $E(K)$ may be adjoined to the family $\{E_i\}_{i \in I}$ to give a larger family of indecomposable injective submodules whose sum is direct. This gives a contradiction. Hence $E' = 0$ and $E = \sum_{i \in I} E_i$ (d.s.), which establishes (b).

It is clear that (b) implies (c) with $\kappa = 1$. Accordingly, we assume (c) and prove that R is Noetherian. Let $\{S_n\}_{n=1}^{\infty}$ be a countably infinite family of simple R-modules. By Theorem 4.1, it is sufficient to prove that $\overset{\infty}{\underset{n=1}{\oplus}} E(S_n)$ is injective. Let I be an infinite set whose cardinality is greater than κ, put $S = \overset{\infty}{\underset{n=1}{\oplus}} S_n$, $T = \underset{i \in I}{\oplus} S$, and let F be an injective envelope of T. Now T is semi-simple and, by Proposition 3.16,

$$S(F) = S(E(T)) = S(T) = T.$$

Also F is injective so, by (c), we can write

$$F = \sum_{j \in J} F_j \quad \text{(d.s.)},$$

where $c(F_j) \leqslant \kappa$. For each j, we write

$$S(F_j) = \sum_{k \in K_j} T_k \quad \text{(d.s.)},$$

where the T_k are simple. Note that $|K_j| \leqslant \kappa$. Hence

$$S(F) = \sum_{j \in J} \sum_{k \in K_j} T_k \quad \text{(d.s.)}$$

and

$$\overset{\infty}{\underset{i \in I}{\oplus}} \underset{n=1}{\oplus} S_n = \sum_{j \in J} \sum_{k \in K_j} T_k \quad \text{(d.s.)}.$$

For each positive integer n, we denote by $J(n)$ the set of all j in J for which there exists $k \in K_j$ such that $T_k \approx S_n$. Each set $J(n)$ is infinite. For suppose that $J(n)$ is finite for some n, say with distinct elements $j_1, j_2, \ldots, j_r$ $(r \geqslant 0)$. It follows from Theorem 3.13 Corollary that

$$|I| \leqslant |K_{j_1}| + |K_{j_2}| + \ldots + |K_{j_r}| \leqslant \kappa + \kappa + \ldots + \kappa \quad (r \text{ terms}).$$

Now $\kappa + \kappa + \ldots + \kappa$ is either finite or κ and so is strictly less than $|I|$. This is a contradiction. But now there is a sequence of distinct elements $l_1, l_2, l_3, \ldots$ such that $l_n \in J(n)$ for all n. For every n, there exists $k_n \in K_{l_n}$ such that $T_{k_n} \approx S_n$. Now $T_{k_n} \subseteq F_{l_n}$ and F_{l_n} is injective, so T_{k_n} has an injective envelope $E(T_{k_n})$ which is a submodule of F_{l_n} (Proposition 2.19). We can now write

$$F_{l_n} = E(T_{k_n}) + F'_{l_n} \quad \text{(d.s.)},$$

where F'_{l_n} is a submodule of F_{l_n}. Then $\overset{\infty}{\underset{n=1}{\sum}} E(T_{k_n})$ (d.s.) is a direct summand of F and so is injective. But $E(T_{k_n}) \approx E(S_n)$ for every n, so $\overset{\infty}{\underset{n=1}{\oplus}} E(S_n)$ is injective. This establishes (a). $\square$

REMARK As an illustration of Theorem 4.4, we note that a ring R is Noetherian if there is a cardinal number κ such that every injective R-module is the direct sum of a family of submodules each of which has cardinality less than or equal to κ. This follows since, for an R-module M,

$$c(M) \leqslant |S(M)| \leqslant |M|.$$

In particular, R *is Noetherian if every injective R-module is the direct sum of a family of finitely generated submodules;* for if M is a finitely generated R-module, say $M = Rm_1 + \dots + Rm_n$, then

$$|M| \leqslant |R \oplus R \oplus \dots \oplus R| = n|R|,$$

and $n|R|$ is either finite or $|R|$.

Theorem 4.4 provides an alternative characterization of Noetherian rings to that given in Theorem 4.1. We now establish a similar characterization of Artinian rings.

THEOREM 4.5 *The following statements are equivalent:*

(a) R *is an Artinian ring;*

(b) *every injective R-module is a direct sum of a family of injective envelopes of simple modules.*

Proof. Assume that R is an Artinian ring. By Theorem 3.25 Corollary, R is also Noetherian, so that every injective R-module is a direct sum of a family of indecomposable injective modules. The statement (b) now follows from Proposition 2.28 Corollary 5.

We now assume (b). Theorem 3.21 says that R is Artinian if every homomorphic image of R is finitely embedded. Consider a left ideal A of R. By (b), we can write

$$E(R/A) = \sum_{i \in I} E(S_i) \quad \text{(d.s.)},$$

where the S_i are simple submodules of $E(R/A)$. Since R/A is singly generated, there exists a finite subset J of I such that

$$R/A \subseteq \sum_{i \in J} E(S_i) \text{ (d.s.)} \subseteq E(R/A)$$

But $\sum_{i \in J} E(S_i)$ is injective, so that

$$E(R/A) = \sum_{i \in J} E(S_i) \quad \text{(d.s.)}$$

and R/A is finitely embedded. This shows that R is Artinian. □

We can use Theorem 4.5 to provide a well-known description of when a commutative Noetherian ring is Artinian, a description not explicitly involving injective modules.

THEOREM 4.6 *Let R be a commutative Noetherian ring. Then the following statements are equivalent:*
(a) *R is Artinian;*
(b) *every prime ideal of R is maximal.*

Proof. Suppose that R is an Artinian ring, and let P be a prime ideal of R. By Theorem 4.5, we can express $E(R/P)$ in the form

$$E(R/P) = \sum_{i \in I} E(S_i) \quad \text{(d.s.)},$$

where the S_i are simple modules. But $E(R/P)$ is indecomposable, so the index set I can contain only one member, i.e.

$$E(R/P) = E(S)$$

for some simple submodule S of $E(R/P)$. Hence there exists a maximal ideal M of R such that

$$E(R/P) \approx E(R/M).$$

Now M is a prime ideal so, by Lemma 2.31 Corollary, $P = M$ and so is maximal.

Now suppose that every prime ideal of R is maximal. By Theorem 4.4, every injective R-module E can be expressed in the form

$$E = \sum_{i \in I} E_i \quad \text{(d.s.)},$$

where the E_i are indecomposable injective modules. By Theorem 2.32 Corollary, there exists for each i a prime ideal P_i such that $E_i \approx E(R/P_i)$. Now P_i is actually maximal, so E_i is an injective envelope of a simple module. It follows from Theorem 4.5 that R is Artinian. □

4.2 The normal decomposition

The next result is similar in nature to Theorem 4.4, or, at any rate, to one of the implications in Theorem 4.4. It has a similar proof.

THEOREM 4.7 *Let M be a Noetherian R-module. Then $E(M)$ is the direct sum of a finite number of indecomposable injective modules.*

Proof. As in the case of Theorem 4.4, $E(M)$ has a family $\{E_i\}_{i \in I}$ of indecomposable injective submodules which is maximal with respect to the property that its sum is direct. Now the submodules $E_i \cap M$ ($i \in I$) are non-zero, and their sum is direct. But M is Noetherian. It follows that the index set I must be finite. Thus $\sum_{i \in I} E_i$ (d.s.) is injective and so is a direct summand of $E(M)$. Write $E(M) = (\sum_{i \in I} E_i) + E'$ (d.s.), where E' is a sub-module of $E(M)$. Then E' is also injective. If $E' \neq 0$, then $E' \cap M$ will be a non-zero submodule of M and so will be Noetherian. Lemma 4.3 now shows that $E' \cap M$ has a submodule K such that $E(K)$ is indecomposable, and we can take $E(K)$ to be a submodule of E'. By adjoining such an envelope to the family $\{E_i\}_{i \in I}$, we obtain a larger family of indecomposable injective submodules of $E(M)$ whose sum is direct. This gives a contradiction and shows that $E(M) = \sum_{i \in I} E_i$ (d.s.). $\square$

We shall now introduce a notion which will be used continually for the rest of this chapter. Let K be a submodule of an R-module M and suppose that $E(M/K)$ is isomorphic to a direct sum of a finite number of indecomposable injective modules, say

$$E(M/K) \approx E_1 \oplus \ldots \oplus E_n.$$

We know from Theorem 3.13 Corollary that, when this occurs, the modules $E_1, \ldots, E_n$ are uniquely determined up to order and isomorphism.

DEFINITION *Let K and M be as in the previous paragraph. We call $E_1, \ldots, E_n$ the complete set of associated indecomposable injective modules or, more simply, the 'complete set of associated* *indecomposable injectives' of K in M; and we say that 'K possesses associated indecomposable injectives in M'.*

Strictly speaking, we should be associating with K not individual indecomposable injective modules but rather isomorphism classes of injectives. However, we shall say that associated indecomposable injectives are defined only up to isomorphism. This will cause no difficulties. We have inserted the word 'complete' to emphasize that there may be isomorphic copies repeated among $E_1, \ldots, E_n$. (This is analogous to the statement that the polynomial $(x-1)^2(x-2)$ has complete set of roots 1, 1, 2.)

We are not associating indecomposable injective modules with every submodule K of M. Indeed, the phrase 'K possesses associated indecomposable injectives in M' will be synonymous with '$E(M/K)$ is a direct sum of a finite number of indecomposable injective modules'.

Here are a number of situations in which K does possess associated indecomposable injectives in M:

(i) M/K *is Noetherian*. This is the statement of Theorem 4.7. This holds in particular when M is Noetherian.

(ii) M/K *is finitely embedded*. Here, the associated indecomposable injectives are all injective envelopes of simple modules.

(iii) M/K *is Artinian*. For by Theorem 3.21 an Artinian module is finitely embedded. This holds in particular when M is Artinian.

(iv) K *is an irreducible submodule of* M. Then K has one associated indecomposable injective in M, namely $E(M/K)$.

(v) $K = M$. It is the empty set of indecomposable injectives that are associated with M in M.

The next result is useful when it comes to recognizing when a given indecomposable injective module is associated with a certain submodule. It follows immediately from Lemma 3.14.

LEMMA 4.8 *Let K be a submodule of an R-module M which possesses associated indecomposable injectives in M, and let E be an indecomposable injective R-module. Then the following statements are equivalent:*

(a) *E is an associated indecomposable injective module of K in M;*

(b) *E is isomorphic to a direct summand of $E(M/K)$.* □

DEFINITION   *Let* $K_1, ..., K_n$ *be submodules of an R-module M. Then the intersection* $K_1 \cap ... \cap K_n$ *is said to be 'irredundant' if, for every* i $(1 \leqslant i \leqslant n)$,

$$K_i \not\supseteq K_1 \cap ... \cap K_{i-1} \cap K_{i+1} \cap ... \cap K_n.$$

In other words, the intersection $K_1 \cap ... \cap K_n$ is irredundant if it is altered by the omission of any one of the K_i. For the purpose of the above definition, we recall that an empty intersection is to mean the whole of M. Also, an empty intersection may be regarded as irredundant. We note that, given submodules $K_1, ..., K_n$ of M, then we can find a subset $\{K_{i_1}, ..., K_{i_r}\}$ of $\{K_1, ..., K_n\}$ such that

(a) $K_{i_1} \cap ... \cap K_{i_r} = K_1 \cap ... \cap K_n$, and
(b) the intersection $K_{i_1} \cap ... \cap K_{i_r}$ is irredundant.

This may be done by omitting appropriate K_is one–by–one.

THEOREM 4.9   *Let* $K_1 \cap ... \cap K_n = K$ *(say) be an irredundant intersection of irreducible submodules of an R-module M. Then*

$$E(M/K) \approx E(M/K_1) \oplus ... \oplus E(M/K_n)$$

and K has the complete set of associated indecomposable injectives $E(M/K_1), ..., E(M/K_n)$ *in M.*

Proof. The mapping

$$M \to E(M/K_1) \oplus ... \oplus E(M/K_n)$$

under which the element m of M maps to $(m + K_1, ..., m + K_n)$ has kernel K, and so induces a monomorphism

$$\phi: M/K \to E(M/K_1) \oplus ... \oplus E(M/K_n).$$

For each i $(1 \leqslant i \leqslant n)$, we denote by ϕ_i the injection mapping

$$\phi_i: E(M/K_i) \to E(M/K_1) \oplus ... \oplus E(M/K_n).$$

By the irredundancy of the intersection $K_1 \cap ... \cap K_n$, there exists, for each i, an element

$$x_i \in K_1 \cap ... \cap K_{i-1} \cap K_{i+1} \cap ... \cap K_n, \; x_i \notin K_i.$$

Then $\phi(x_i + K) = \phi_i(x_i + K_i)$ is a non-zero element of

$$\phi(M/K) \cap \phi_i(M/K_i).$$

Now $E(M/K_i)$, and so also $\phi_i(E(M/K_i))$, is indecomposable. It follows that $\phi_i(E(M/K_i))$ is an injective envelope of

$$\phi(M/K) \cap \phi_i(M/K_i).$$

Hence, by Proposition 2.23,

$$\bigoplus_{i=1}^{n} E(M/K_i) = \sum_{i=1}^{n} \phi_i(E(M/K_i)) \quad \text{(d.s.)}$$

is an injective envelope of

$$\sum_{i=1}^{n} \phi(M/K) \cap \phi_i(M/K_i) \quad \text{(d.s.)},$$

and hence also of $\phi(M/K)$. The result follows from this. $\square$

COROLLARY 1 *Let K be a submodule of an R-module M and suppose that*
$$K_1 \cap \ldots \cap K_n = K = K_1' \cap \ldots \cap K_p'$$

are expressions for K as an irredundant intersection of irreducible submodules of M. Then $n = p$ and there is a one–one correspondence between the $E(M/K_i)$ and the $E(M/K_j')$ such that corresponding modules are isomorphic. $\square$

The result that $n = p$ is a special case of the celebrated *Kurosh–Ore theorem* for lattices.

COROLLARY 2 *Let K be a submodule of an R-module M which can be expressed as the intersection of a finite number of irreducible submodules of M. Then K possesses associated indecomposable injectives in M (i.e. $E(M/K)$ is the direct sum of a finite family of indecomposable injective modules).*

Proof. This follows because, beginning with an expression for K as the intersection of a finite number of irreducible submodules, we can omit certain of the terms one–by–one to obtain an expression for K as an irredundant intersection of irreducible submodules. $\square$

Theorem 4.9 Corollary 2 describes a situation in which associated indecomposable injectives exist. In fact, this is the only situation in which they exist, as the next result, called the *first decomposition theorem*, shows.

THEOREM 4.10   *Let K be a submodule of an R-module M. Then the following statements are equivalent:*

(a) *$E(M/K)$ is the direct sum of a finite number of indecomposable injective modules;*

(b) *K is the intersection of a finite number of irreducible submodules of M.*

Proof. It remains to show that (a) implies (b). Suppose that

$$E(M/K) = E_1 + \ldots + E_n \quad \text{(d.s.)},$$

where $E_1, \ldots, E_n$ are indecomposable injective R-modules. For each i, consider the combined homomorphism

$$\phi_i \colon M \to M/K \xrightarrow{\text{inc}} E(M/K) \xrightarrow{\text{proj}} E_i,$$

where $M \to M/K$ is the natural mapping. Put $K_i = \operatorname{Ker} \phi_i$. Then

$$K = K_1 \cap \ldots \cap K_n. \tag{4.2.1}$$

Since $E_i \cap (M/K) \neq 0$, it follows that $K_i \neq M$. Further,

$$M/K_i \approx \phi_i(M) \subseteq E_i,$$

so that $$E(M/K_i) \approx E_i$$

because E_i is an indecomposable injective module. It follows that K_i is irreducible. $\square$

REMARK   The intersection (4.2.1) is actually irredundant. For, if it were not, we could obtain from it an irredundant intersection with fewer terms, and by Theorem 4.9 $E(M/K)$ would be isomorphic to a direct sum of fewer than n indecomposable injective modules.

COROLLARY   *Let M be an R-module which is either Noetherian or Artinian and let K be a submodule of M. Then K is the intersection of a finite number of irreducible submodules of M.*

Proof. See the remarks preceding Lemma 4.8. $\square$

DEFINITION   *Let K be a submodule of an R-module M and suppose that K has the complete set of associated indecomposable injectives $E_1, \ldots, E_n$, where $n \geq 1$. Suppose further that $E_1, \ldots, E_n$ are all isomorphic to each other. Then K is said to be 'isotopic'; more explicitly, if $E_1 \approx \ldots \approx E_n \approx E$, then K is said to be 'E-isotopic'.*

Thus, to say that K is E-isotopic in M, where E is an indecomposable injective module, is to say that

$$E(M/K) \approx E \oplus \ldots \oplus E,$$

where there are only finitely many Es in the direct sum. Alternatively, it amounts to saying that K can be expressed in the form
$$K = K_1 \cap \ldots \cap K_r,$$
where each K_i is an irreducible submodule of M and

$$E(M/K_1) \approx \ldots \approx E(M/K_r) \approx E.$$

It is immaterial whether we require that the intersection be irredundant.

If K is an irreducible submodule of the R-module M, then K is $E(M/K)$-isotopic. We note that an isotopic submodule of a module is a proper submodule by definition.

The next result is obvious from the above remarks.

PROPOSITION 4.11   *Let $K_1, \ldots, K_n$ be E-isotopic submodules of an R-module M. Then $K_1 \cap \ldots \cap K_n$ is an E-isotopic submodule of M.*□

DEFINITION   *Let K be a submodule of an R-module M. Then an intersection $K = K_1 \cap \ldots \cap K_n$ is called a 'normal decomposition' of K in M if*

(a) the intersection is irredundant,

(b) for each i, K_i is E_i-isotopic for some indecomposable injective module E_i,

(c) the E_i are non-isomorphic.

Let K be a submodule of an R-module M, and suppose that

$$K = K_1 \cap \ldots \cap K_n. \tag{4.2.2}$$

is an intersection of irreducible submodules of M. Then, for each i, K_i is $E(M/K_i)$-isotopic. By grouping together those K_is whose associated indecomposable injectives are isomorphic, and using Proposition 4.11, we obtain an expression for K as the intersection of a finite number of isotopic submodules whose associated indecomposable injectives are non-isomorphic. This may then be made irredundant by deleting appropriate terms one–by–one from the intersection to obtain a normal decomposition of K in M.

We now go in the reverse direction, keeping a careful watch on the indecomposable injective modules in the process. This time we let (4.2.2) be a *normal* decomposition of K in M, where K_i is E_i-isotopic for $1 \leqslant i \leqslant n$. Then, for $1 \leqslant i \leqslant n$, we can write

$$K_i = K_{i1} \cap \ldots \cap K_{ip_i},$$

where $E(M/K_{ij}) \approx E_i$ for $1 \leqslant j \leqslant p_i$. This enables us to express K as an intersection of irreducible submodules of M. Let us examine this intersection more closely. We can obtain from it an irredundant intersection of irreducible submodules by the systematic deletion of certain of the K_{ij}. For $1 \leqslant i \leqslant n$, we denote by K_i' the intersection of those K_{ij}s, where $1 \leqslant j \leqslant p_i$, which remain. Then $K_i' \supseteq K_i$ and

$$K = K_1' \cap \ldots \cap K_n'.$$

Suppose that all the K_{1j}s for $1 \leqslant j \leqslant p_1$ have been deleted, so that $K_1' = M$. Then

$$K_1 \supseteq K_1 \cap \ldots \cap K_n = K_1' \cap \ldots \cap K_n' = K_2' \cap \ldots \cap K_n' \supseteq K_2 \cap \ldots \cap K_n.$$

This is imposssible because (4.2.2) was assumed to be a normal decomposition of K in M. Thus not all the K_{1j}s have been deleted, nor have all the K_{2j}s, nor all the K_{3j}s and so on. Now what can be deduced from this? Let us look at the complete set of associated indecomposable injectives of K in M. By Theorem 4.9, we shall have a positive number of copies of E_1, a positive number of copies of E_2 and so on. In other words, $E_1, \ldots, E_n$ *are the associated indecomposable injectives of K in M where now we have NOT repeated isomorphic copies the appropriate number of times.* In this situation we call $E_1, \ldots, E_n$ the *reduced* set of indecomposable injectives of K in M. (In an analogous way, we would say that the polynomial $(x-1)^2 (x-2)$ has reduced set of roots 1, 2.)

If we allow Theorem 4.10 to interact with the above remarks, we obtain the *second decomposition theorem*.

Theorem 4.12   *Let K be a submodule of an R-module M. Then the following statements are equivalent:*

(a) *$E(M/K)$ is the direct sum of a finite number of indecomposable injective modules;*

(b) *K has a normal decomposition in M.*

Further, if $K = K_1 \cap \ldots \cap K_n$ is a normal decomposition of K in M, where K_i is E_i-isotopic, then K has $E_1, \ldots, E_n$ as its reduced set of associated indecomposable injectives in M. □

COROLLARY 1 *Let K be a submodule of an R-module M and suppose that*
$$K_1 \cap \ldots \cap K_n = K = K_1' \cap \ldots \cap K_p'$$
are normal decompositions of K in M, where K_i is E_i-isotopic and K_j' is E_j'-isotopic. Then $n = p$ and there is a one–one correspondence between the E_i and E_j' such that corresponding modules are isomorphic. □

As in the case of Theorem 4.10, we can obtain a further corollary to Theorem 4.12.

COROLLARY 2 *Let M be an R-module which is either Noetherian or Artinian and let K be a submodule of M. Then K has a normal decomposition in M.* □

If $K = K_1 \cap \ldots \cap K_n$ is a normal decomposition of the submodule K of M, where K_i is E_i-isotopic, we refer to K_i as the *E_i-isotopic component* in this normal decomposition. It is conspicuous that no claim has been made concerning uniqueness of the separate components of a normal decomposition; indeed, it is not true in general that these are unique. We can however provide a partial result in this direction.

We consider the submodule K of M with reduced set of associated indecomposable injectives $E_1, \ldots, E_n$. Consider a nonempty subset Ω of $\{E_1, \ldots, E_n\}$. We say that Ω is an *isolated set of associated indecomposable injectives of K in M* if

$$\text{Hom}_R(F, E) = 0 \text{ whenever } E \in \Omega \quad \text{and} \quad F \in \{E_1, \ldots, E_n\}, F \notin \Omega.$$

An associated indecomposable injective E of K in M such that $\{E\}$ is an isolated set is said to be an *isolated associated indecomposable injective of K in M*. The whole reduced set of indecomposable injectives of K in M is isolated, but this is not very interesting as can be seen from the use to which these isolated sets are put in the next theorem.

THEOREM 4.13 *Let K be a submodule of an R-module M and suppose that K has a normal decomposition in M. Let*

$$K_1 \cap \ldots \cap K_n = K = K_1' \cap \ldots \cap K_n'$$

*be two such normal decompositions, where K_i and K'_i are E_i-isotopic
for $1 \leqslant i \leqslant n$. Suppose that $E_{i_1}, \ldots, E_{i_r}$, where $1 \leqslant i_1 < \ldots < i_r \leqslant n$,
form an isolated set of associated indecomposable injectives of K
in M. Then*
$$K_{i_1} \cap \ldots \cap K_{i_r} = K'_{i_1} \cap \ldots \cap K'_{i_r}.$$

*In particular, if E is an isolated associated indecomposable injective
of K in M, then the E-isotopic component of K in M is independent
of the particular normal decomposition.*

REMARK   This uniqueness result has about it a hollow tone,
because it is not clear that there are isolated sets of associated
indecomposable injectives other than the whole set. In particular,
it is not clear that K possesses isolated associated indecomposable
injectives in M. We shall be able to lend some substance to the
result in the next section. We shall see, for example, that, when
M is a Noetherian or an Artinian module over a commutative
Noetherian ring, then every proper submodule of M possesses
isolated associated indecomposable injectives.

Proof. Consider the normal decomposition
$$K = K_1 \cap \ldots \cap K_n$$
of K in M, where K_i is E_i-isotopic for $1 \leqslant i \leqslant n$. To simplify the
notation, we suppose that $\{E_1, \ldots, E_r\}$ is an isolated set of associated indecomposable injectives of K in M. We define the
submodule L of M, $L \supseteq K$, by
$$L/K = \bigcap_{i=1}^{r} \bigcap_{\phi_i} \{\mathrm{Ker}\,\phi_i : \phi_i \in \mathrm{Hom}_R\,(M/K, E_i)\}.$$
We shall show that
$$L/K = (K_1 \cap \ldots \cap K_r)/K.$$
This in turn will show that $K_1 \cap \ldots \cap K_r = L$ and will establish
that $K_1 \cap \ldots \cap K_r$ is independent of the particular normal
decomposition.

Consider a particular i, $1 \leqslant i \leqslant r$. Now K_i is E_i-isotopic, so
there exists an isomorphism
$$\alpha_i \colon E(M/K_i) \to E_i \oplus \ldots \oplus E_i,$$
where there are finitely many terms, say r_i, in the direct sum.

Let $1 \leqslant j \leqslant r_i$. We denote by π_{ij} the jth projection mapping of this direct sum and consider the composite mapping

$$\alpha_{ij}: M/K \overset{\eta_i}{\to} M/K_i \overset{\text{inc}}{\to} E(M/K_i) \overset{\alpha_i}{\to} E_i \oplus \ldots \oplus E_i \overset{\pi_{ij}}{\to} E_i,$$

where η_i is induced by the identity mapping of M. Then $\alpha_{ij} \in \mathrm{Hom}_R(M/K, E_i)$ and

$$\bigcap_{j=1}^{r_i} \mathrm{Ker}\, \alpha_{ij} = K_i/K.$$

Thus $\qquad \displaystyle\bigcap_{i=1}^{r} \bigcap_{j=1}^{r_i} \mathrm{Ker}\, \alpha_{ij} = (K_1 \cap \ldots \cap K_r)/K,$

so that $\qquad\qquad (K_1 \cap \ldots \cap K_r)/K \supseteq L/K.$

It remains to prove that $(K_1 \cap \ldots \cap K_r)/K \subseteq L/K$. Let

$$\eta: M/K \to M/(K_1 \cap \ldots \cap K_r)$$

be the mapping induced by the identity mapping of M and consider $\phi \in \mathrm{Hom}_R(M/K, E_i)$ for some i, $1 \leqslant i \leqslant r$. If we can show that there exists an R-homomorphism

$$\beta: M/(K_1 \cap \ldots \cap K_r) \to E_i$$

such that $\phi = \beta\eta$, then $\mathrm{Ker}\, \phi \supseteq \mathrm{Ker}\, \eta = (K_1 \cap \ldots \cap K_r)/K$ and the result will follow. The exact sequence

$$0 \to (K_1 \cap \ldots \cap K_r)/K \overset{\text{inc}}{\to} M/K \overset{\eta}{\to} M/(K_1 \cap \ldots \cap K_r) \to 0$$

gives rise to the exact sequence

$$0 \to \mathrm{Hom}_R(M/(K_1 \cap \ldots \cap K_r), E_i) \overset{\zeta}{\to} \mathrm{Hom}_R(M/K, E_i)$$
$$\to \mathrm{Hom}_R((K_1 \cap \ldots \cap K_r)/K, E_i)$$

(see Proposition 1.26). What we have to show is that the mapping ζ is an isomorphism. In other words, we need to prove that

$$\mathrm{Hom}_R((K_1 \cap \ldots \cap K_r)/K, E_i) = 0 \text{ for } 1 \leqslant i \leqslant r.$$

Consider a particular i, $1 \leqslant i \leqslant r$. The mapping

$$\psi: (K_1 \cap \ldots \cap K_r)/K \to E(M/K_{r+1}) \oplus \ldots \oplus E(M/K_n)$$

given by $\qquad \psi(x + K) = (x + K_{r+1}, \ldots, x + K_n),$

where $x \in K_1 \cap \ldots \cap K_r$, is a monomorphism. Since E_i is injective, this gives rise to an epimorphism

$$\mathrm{Hom}_R(E(M/K_{r+1}) \oplus \ldots \oplus E(M/K_n), E_i)$$
$$\to \mathrm{Hom}_R((K_1 \cap \ldots \cap K_r)/K, E_i).$$

But
$$E(M/K_j) \approx E_j \oplus \ldots \oplus E_j$$

for $r+1 \leqslant j \leqslant n$, where there are finitely many summands, say r_j. Hence, by Proposition 1.24, there is an epimorphism of the form

$$\mathop{\oplus}_{j=r+1}^{n} \mathop{\oplus}_{k=1}^{r_j} (\mathrm{Hom}_R (E_j, E_i)) \to \mathrm{Hom}_R((K_1 \cap \ldots \cap K_r)/K, E_i).$$

But $\{E_1, \ldots, E_r\}$ is an isolated set of associated indecomposable injectives, so that

$$\mathrm{Hom}_R (E_j, E_i) = 0 \text{ for } r+1 \leqslant j \leqslant n.$$

It follows that

$$\mathrm{Hom}_R ((K_1 \cap \ldots \cap K_r)/K, E_i) = 0 \text{ for } 1 \leqslant i \leqslant r,$$

as required.□

4.3 The Lasker–Noether decomposition

In this section we shall examine the normal decomposition in a restricted situation. We shall consider a Noetherian module M over a commutative ring. By Theorem 4.12 Corollary 2, every submodule of M has a normal decomposition. We shall describe this decomposition in terms which do not explicitly involve injective modules. Some readers may be familiar with the Lasker–Noether decomposition of a submodule of a module over a commutative ring. (See, for example, DGN *Lessons* Chapter 2.) It is precisely this decomposition that we shall be describing. Thus the normal decomposition reduces to the Lasker–Noether decomposition in our restricted situation. *Throughout this section, R will denote a commutative ring.*

Let K be a submodule of an R-module M which has associated indecomposable injectives in M; let $E_1, \ldots, E_n$ be the reduced set of indecomposable injectives of K in M. Suppose further that there exist prime ideals $P_1, \ldots, P_n$ such that $E_i \approx E(R/P_i)$ for $1 \leqslant i \leqslant n$. Even though $E_1, \ldots, E_n$ are determined by M and K

only up to isomorphism, Lemma 2.31 Corollary shows that not even this degree of ambiguity occurs with $P_1, ..., P_n$; they are uniquely determined by M and K. Also, as no isomorphisms occur between different E_i, $P_1, ..., P_n$ are all different.

DEFINITION *Let the situation be as just described. We call $P_1, ..., P_n$ the 'associated prime ideals of K in M' and we say that 'K has associated prime ideals in M'.*
The phrase 'in M' will often be omitted if no ambiguity will thereby arise. For example, we shall refer to 'the associated prime ideals of an ideal' (if they exist), when it is always understood that this is in the ring considered as a module over itself.

EXAMPLES
1. The R-module M has the empty set of prime ideals as its associated prime ideals in itself, and is the only submodule with this property.
2. A prime ideal P of R has exactly one associated prime ideal, namely P itself.
3. Suppose that M/K is finitely embedded; for example, M might be Artinian. We know that K has associated indecomposable injectives in M and that these are injective envelopes of simple modules. It follows that K has associated prime ideals in M and that these are all maximal ideals. The converse is also true, namely that, if K has associated prime ideals in M which are all maximal ideals, then M/K is finitely embedded.
A further situation in which associated prime ideals exist will be seen in Theorem 4.15.

LEMMA 4.14 *Let M be a Noetherian R-module. Then any direct summand of $E(M)$ is N-injective.*
Proof. Consider a non-zero direct summand E' of $E(M)$. Then E' is injective. Also, $E' \cap M \neq 0$ and, since M is Noetherian, so is $E' \cap M$. It follows that E' is N-injective. $\square$

THEOREM 4.15 *Let M be a Noetherian R-module and let K be a submodule of M. Then K has associated prime ideals in M and these are N-prime ideals.*
Proof. We know that K has associated indecomposable injectives in M and these will be direct summands of $E(M/K)$.

Since also M/K is Noetherian, it follows from Lemma 4.14 that these indecomposable injectives are N-injective. The result now follows by Theorem 2.32. $\square$

THEOREM 4.16   *Let K be a submodule of an R-module M which has associated prime ideals in M and let P be a prime ideal of R. Then the following statements are equivalent:*

(a) *P is an associated prime ideal of K in M;*

(b) *there exists an element m of M such that $P = K:m$.*

Proof. Suppose that P is an associated prime ideal of K in M. Then $E(R/P)$ is isomorphic to a direct summand of $E(M/K)$, so that $E(M/K)$ has a submodule E' (say) isomorphic to R/P. Then $(M/K) \cap E' \neq 0$. Let x be a non-zero element of $(M/K) \cap E'$, so that x is a coset $m + K$, where $m \in M$. Then $K:m = 0:x = P$ and (a) implies (b).

Now suppose that there is an element m of M such that $P = K:m$, and denote by $\overline{m}$ the image of m under the natural mapping $M \to M/K$. Then $P = 0:\overline{m}$, so that $R\overline{m} \approx R/P$. It follows from Proposition 2.22 that $E(R/P)$ is isomorphic to a direct summand of $E(M/K)$. Since $E(R/P)$ is indecomposable, Lemma 4.8 shows that P is an associated prime ideal of K in M. Thus (b) implies (a). $\square$

DEFINITION   *Let K be a submodule of an R-module M. Then K is said to be a 'primary submodule of M' if (a) K is proper, and (b) whenever $rm \in K$ ($r \in R$, $m \in M$), then either $m \in K$ or there is a positive integer n such that $r^n M \subseteq K$.*

Let K be a primary submodule of an R-module M, and denote by P the set of all elements r of R such that $r^n M \subseteq K$ for some positive integer n. Then (b) can be reworded as follows:

(b)′ whenever $rm \in K$ ($r \in R$, $m \in M$), then either $r \in P$ or $m \in K$.

LEMMA 4.17   *P is a prime ideal of R.*

Proof. We shall allow the reader to prove that P is a proper ideal of R. Let the elements α, β of R be such that $\alpha\beta \in P$, and suppose that $\beta \notin P$. Then there is a positive integer n such that $(\alpha\beta)^n M \subseteq K$. However, $\beta^n M \nsubseteq K$, so there exists $m \in M$ such that $\beta^n m \notin K$. But $\alpha^n \beta^n m \in K$, so there is a positive integer r such that $(\alpha^n)^r M \subseteq K$. This shows that $\alpha \in P$, which in turn shows that P is prime. $\square$

DEFINITION   *Let K be a primary submodule of an R-module M, and let P be the prime ideal defined above. We describe this by saying that K is 'P-primary'.*

If K is a P-primary submodule of M, then

$$K:M \subseteq P.$$

In particular, if Q is a P-primary ideal of R, then

$$Q \subseteq P.$$

LEMMA 4.18   *Let K be a P-primary submodule of an R-module M, and suppose that the ring R is Noetherian. Then $P^n M \subseteq K$ for some positive integer n.*

Proof. Since R is Noetherian, P is finitely generated, say $P = Rr_1 + \ldots + Rr_s$. There are positive integers $n_1, \ldots, n_s$ such that $r_i^{n_i} M \subseteq K$ for each i. Put $n = n_1 + \ldots + n_s$. Then, if $x_1, \ldots, x_n$ are arbitrary elements of P, $(x_1 x_2 \ldots x_n) M \subseteq K$. It follows that $P^n M \subseteq K$. □

Before we consider the connection between the notions of isotopic and primary submodules, we shall make a simplification in terminology. Let P be a prime ideal of R. Then the injective module $E(R/P)$ is an indecomposable injective module.

DEFINITION   *An $E(R/P)$-isotopic submodule of an R-module is said to be 'P-isotopic'.*

Thus a submodule K of an R-module M is P-isotopic if and only if it has exactly one associated prime ideal, namely P. For example, $E(R/P)$ is P-isotopic. In fact, K is P-isotopic if and only if there is an isomorphism

$$E(M/K) \approx E(R/P) \oplus \ldots \oplus E(R/P),$$

where there is a finite positive number of summands on the right hand side. We note that, if a submodule is both P-isotopic and P'-isotopic, where P' is also a prime ideal of R, then $P = P'$. Also, a P-isotopic submodule of M is a proper submodule of M.

THEOREM 4.19   *Let M be a Noetherian R-module, let K be a submodule of M and let P be a prime ideal of R. Then the following statements are equivalent:*

   (a) *K is a P-primary submodule of M;*

   (b) *K is a P-isotopic submodule of M.*

Proof. Suppose that K is a P-primary submodule of M. Then K is a proper submodule of M. Also, K has associated prime ideals in M, since M is Noetherian (Theorem 4.15). Let P' be such an associated prime ideal, which exists because K is a proper submodule of M. By Theorem 4.16, there is an element m of M such that $P' = K:m$. Now $m \notin K$ because $P' \neq R$, so, by $(b)'$ above, every element of P' must belong to P, i.e. $P' \subseteq P$. But, if $r \in P$, then there is a positive integer n such that $r^n M \subseteq K$. Then $r^n \in K:m = P'$, whence $r \in P'$ because P' is prime. Hence $P' = P$, so that P is the one and only associated prime ideal of K in M and K is P-isotopic.

Now suppose that K is a P-isotopic submodule of M. Then K is a proper submodule of M. Further,

$$E(M/K) \approx E(R/P) \oplus \ldots \oplus E(R/P)$$
$$= E((R/P) \oplus \ldots \oplus (R/P)),$$

where there are finitely many summands, say l. Note that $l > 0$. Hence $E(M/K)$ has a submodule $L \approx (R/P) \oplus \ldots \oplus (R/P)$ (l terms) such that $E(L) = E(M/K)$.

Suppose that $rm \in K$ ($r \in R, m \in M$) and that $m \notin K$. Denote by $\overline{m}$ the image of m under the natural mapping $M \to M/K$. There exists an element s of R such that $s\overline{m}$ is a non-zero element of L. By considering the image of $s\overline{m}$ in $(R/P) \oplus \ldots \oplus (R/P)$ under the isomorphism that exists between L and this module, we see that $0:s\overline{m} = P$. Hence $r \in 0:\overline{m} \subseteq 0:s\overline{m} = P$.

We denote by Q the set of all elements r of R such that $r^n M \subseteq K$ for some positive integer n. If we can show that $Q = P$, then we have shown that K is a P-primary submodule of M. Suppose that $r \in Q$. Then $r^n M \subseteq K$ for some positive integer n. Since $K \neq M$, the previous paragraph shows that $r^n \in P$, whence also $r \in P$.

Now suppose that $r \in P$ and, for each positive integer n, put $A_n = 0:_{M/K} r^n$. Then each A_n is a submodule of M/K and

$$A_1 \subseteq A_2 \subseteq A_3 \subseteq \ldots.$$

Since M/K is Noetherian, there exists a positive integer n_0 such that $A_n = A_{n_0}$ whenever $n \geqslant n_0$. Suppose that $A_{n_0} \neq M/K$. Then $r^{n_0}(M/K)$ is a non-zero submodule of M/K, so that

$$(r^{n_0}(M/K)) \cap L \neq 0.$$

Consider a non-zero element of $(r^{n_0}(M/K)) \cap L$. This can be written as $r^{n_0}\eta$ for some $\eta \in M/K$. Then $\eta \notin A_{n_0}$ but $\eta \in A_{n_0+1}$, since an element of L is annihilated by every element of P. This is impossible. It follows that $A_{n_0} = M/K$, so that $r^{n_0}M \subseteq K$ and $r \in Q$. Thus $Q = P$. $\square$

COROLLARY *Let R be a Noetherian ring, let I be an ideal of R and let P be a prime ideal of R. Then the following statements are equivalent:*

(a) *I is a P-primary ideal of R;*

(b) *I is a P-isotopic ideal of R.* $\square$

Let M be a Noetherian R-module and let K be a submodule of M. By Theorem 4.15, K has associated prime ideals in M; let these be $P_1, ..., P_n$. Then K has a normal decomposition

$$K = K_1 \cap ... \cap K_n$$

in M, where K_i is P_i-isotopic for $1 \leqslant i \leqslant n$. Theorem 4.19 gives that, instead of describing K_i as P_i-isotopic, we may describe it as P_i-primary; the two designations are equivalent. A normal decomposition of K in M is thus an expression of the form

$$K = K_1 \cap ... \cap K_n,$$

where

(a) the intersection is irredundant,

(b) for each i, K_i is P_i-primary for some prime ideal P_i, and

(c) the P_i are distinct.

This is usually referred to as the *Lasker–Noether decomposition* of K in M. The set of prime ideals $\{P_1, ..., P_n\}$ is unique; they are the associated prime ideals of K in M. Thus *the normal decomposition reduces to the Lasker–Noether decomposition when the ring is commutative and M is Noetherian.*

We aim next to examine Theorem 4.13 in the setting of the Lasker–Noether decomposition. We first prove two propositions.

PROPOSITION 4.20 *Let Q be an ideal and P a prime ideal of R. Then the following statements are equivalent:*

(a) *Q is irreducible and P-isotopic;*

(b) *Q is of the form $0:e$, where e is a non-zero element of $E(R/P)$.*

Proof. Assume that Q is irreducible and P-isotopic. Then $E(R/Q)$ is indecomposable, and

$$E(R/Q) \approx E(R/P).$$

The element $1 + Q$ of R/Q has annihilator Q. If e is the element in $E(R/P)$ corresponding to $1 + Q$, then $0 : e = Q$. Note that $e \neq 0$ because $Q \neq R$. Thus (a) implies (b).

Now suppose that $Q = 0 : e$, where e is a non-zero element of $E(R/P)$. The mapping $R \to E(R/P)$ defined by multiplication by e induces a monomorphism $R/(0 : e) \to E(R/P)$. Since $E(R/P)$ is indecomposable, $E(R/P) \approx E(R/(0 : e))$, so that $0 : e$ is irreducible and P-isotopic. Thus (b) implies (a). □

PROPOSITION 4.21 *Let R be a Noetherian ring and P_1, P_2 prime ideals of R. Then the following statements are equivalent:*

   (a) $P_2 \subseteq P_1$;

   (b) $\mathrm{Hom}_R(E(R/P_2), E(R/P_1)) \neq 0$.

Proof. Suppose first that $P_2 \subseteq P_1$. Then the identity mapping of R induces a homomorphism $R/P_2 \to R/P_1$, and the diagram Fig. 4.3, with the obvious mappings, can be completed by a homomorphism $\phi : E(R/P_2) \to E(R/P_1)$. Moreover, ϕ is not the zero mapping. Thus (a) implies (b).

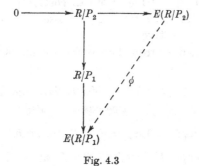

Fig. 4.3

Now assume (b), and let $\phi : E(R/P_2) \to E(R/P_1)$ be a non-zero homomorphism. Let e be an element of $E(R/P_2)$ such that $\phi(e) \neq 0$. By Lemma 2.31, $0 : e \subseteq 0 : \phi(e) \subseteq P_1$. By Proposition 4.20, $0 : e$ is a P_2-isotopic ideal of R. Theorem 4.19 Corollary shows that $0 : e$ is also P_2-primary. Let $r \in P_2$. Then there exists a positive integer n such that $r^n \in 0 : e \subseteq P_1$. But P_1 is prime, so $r \in P_1$. This shows that $P_2 \subseteq P_1$, and (b) implies (a). □

Now let R be a Noetherian ring and let K be a submodule of an R-module M which has associated prime ideals in M. Denote the associated prime ideals of K in M by $P_1, \ldots, P_n$; then K has reduced set of associated indecomposable injectives $E(R/P_1), \ldots, E(R/P_n)$. Suppose that $\{E(R/P_1), \ldots, E(R/P_r)\}$ is an isolated set and consider an associated prime ideal P other than $P_1, \ldots, P_r$. If $P \subseteq P_i$ for some i, $1 \leqslant i \leqslant r$, then by Proposition 4.21

$$\mathrm{Hom}_R\,(E(R/P),\,E(R/P_i)) \,\neq\, 0,$$

which is not so. Thus $P \nsubseteq P_i$ for $1 \leqslant i \leqslant r$. Conversely, if the set $\{P_1, \ldots, P_r\}$ has the property that none of $P_{r+1}, \ldots, P_n$ is contained in any of $P_1, \ldots, P_r$, then $\{E(R/P_1), \ldots, E(R/P_r)\}$ is an isolated set. When $\{P_1, \ldots, P_r\}$ has this property, we say that it is an *isolated set of associated prime ideals of K in M*. Thus an isolated set of associated prime ideals corresponds to an isolated set of associated indecomposable injectives. An associated prime ideal P such that $\{P\}$ is an isolated set is said to be an *isolated associated prime ideal of K in M*. This corresponds to an isolated associated indecomposable injective. An isolated associated prime ideal of K in M is just a minimal member of the set of associated prime ideals. Such a minimal member can certainly be found when $K \neq M$, so in this situation K possesses isolated associated indecomposable injectives.

We can now deduce from Theorem 4.13 the following uniqueness theorem:

THEOREM 4.22 *Let R be a Noetherian ring, let K be a submodule of an R-module M which has associated prime ideals in M and let*

$$K_1 \cap \ldots \cap K_n = K = K_1' \cap \ldots \cap K_n'$$

be normal decompositions of K in M, where K_i and K_i' are P_i-isotopic for $1 \leqslant i \leqslant n$. Suppose that $P_{i_1}, \ldots, P_{i_r}$, where $1 \leqslant i_1 < \ldots < i_r \leqslant n$, form an isolated set of associated prime ideals of K in M. Then

$$K_{i_1} \cap \ldots \cap K_{i_r} = K_{i_1}' \cap \ldots \cap K_{i_r}'.$$

In particular, if P is an isolated associated prime ideal of K in M, then the P-isotopic component of K in M is independent of the particular normal decomposition. $\square$

When M is Noetherian, we can replace 'P_i-isotopic' by 'P_i-primary' in the statement of the theorem, and the result becomes the uniqueness theorem for the Lasker–Noether decomposition.

PROPOSITION 4.23   *Let R be a Noetherian ring and P a prime ideal of R. Put $E = E(R/P)$ and $A_n = 0:_E P^n$ $(n = 1, 2, 3, ...)$. Then*

$$E = \bigcup_{n=1}^{\infty} A_n.$$

Further, when P is a maximal ideal of R, then

$$0:A_n = P^n$$

and
$$\operatorname{Ann}_R E = \bigcap_{n=1}^{\infty} P^n.$$

Proof. Let e be a non-zero element of E. By Proposition 4.20 and Theorem 4.19 Corollary, $0:e$ is a P-primary ideal of R, so, by Lemma 4.18, there is a positive integer n such that $P^n \subseteq 0:e$, whence $e \in A_n$. It follows that $E = \bigcup_{n=1}^{\infty} A_n$.

Now suppose that P is a maximal ideal of R and consider a particular positive integer n. Since $P^n A_n = 0$, we already know that $P^n \subseteq 0:A_n$, so we need to establish the opposite inclusion. By Theorem 4.10 Corollary, we can write

$$P^n = Q_1 \cap ... \cap Q_k,$$

where $Q_1, ..., Q_k$ are irreducible ideals of R. Let $1 \leqslant i \leqslant k$. Then $E(R/Q_i)$ is indecomposable, so, by Theorem 2.32 Corollary, $E(R/Q_i) \approx E(R/P_i)$ for some prime ideal P_i. Thus Q_i is P_i-isotopic, and so P_i-primary. Hence $P^n \subseteq Q_i \subseteq P_i$, so $P \subseteq P_i$. But P is maximal, so $P = P_i$. Thus Q_i is irreducible and P-isotopic, and Proposition 4.20 shows that $Q_i = 0:e_i$ for some non-zero element e_i of E. Hence $P^n e_i = 0$, so that $e_i \in A_n$. This is true for each i, so that
$$0:A_n \subseteq (0:e_1) \cap ... \cap (0:e_k)$$
$$= Q_1 \cap ... \cap Q_k$$
$$= P^n.$$

Finally, with P a maximal ideal of R,

$$\operatorname{Ann}_R E = \bigcap_{n=1}^{\infty} (0:A_n) = \bigcap_{n=1}^{\infty} P^n. \qquad \square$$

COROLLARY 1 *Let R be a Noetherian domain and M a maximal ideal of R. Then*

$$\bigcap_{n=1}^{\infty} M^n = 0.$$

Proof. We have $\bigcap_{n=1}^{\infty} M^n = \operatorname{Ann}_R E(R/M)$, and this is zero by Proposition 2.26 Corollary 1. ☐

DEFINITION *A commutative Noetherian quasi-local ring is called a 'local ring'.*
Thus a commutative Noetherian ring is local if it has a unique maximal ideal.

COROLLARY 2 *Let R be a local ring with maximal ideal M. Then*

$$\bigcap_{n=1}^{\infty} M^n = 0.$$

Proof. Use Proposition 2.26 Corollary 2. ☐

Proposition 4.23 enables us to establish the characterization, promised in Chapter 2, of a Dedekind ring in terms of injective modules. We need a lemma.

LEMMA 4.24 *Let M be a maximal ideal of R and let*

$$E = E(R/M).$$

Then $$R/M = 0 :_E M.$$

Proof. Certainly $R/M \subseteq 0 :_E M$. But $0 :_E M$ may be regarded as a vector space over the field R/M, and R/M is a subspace of $0 :_E M$. Thus R/M is a direct summand of $0 :_E M$, say

$$0 :_E M = (R/M) + F \quad \text{(d.s.)},$$

where F is an (R/M)-module. But F is also an R-module, and $(R/M) \cap F = 0$. However, E is indecomposable so, by Proposition 2.28, $F = 0$. ☐

THEOREM 4.25 *Let R be a commutative domain. Then the following statements are equivalent:*
(a) *R is a Dedekind domain;*
(b) *every divisible R-module is injective.*

Proof. We proved in Proposition 2.10 that (a) implies (b), so we assume (b) and seek to deduce (a). By Lemma 2.5, every direct sum of divisible modules is divisible, so in our case every direct sum of injective modules is injective. Theorem 4.1 now gives that R is a Noetherian ring.

By Lemma 2.9, we must prove that every non-zero integral ideal of R is invertible. If R is a field, this is certainly true, so we shall suppose that R is not a field. We consider first a maximal ideal M of R, and denote by K the field of fractions of R. As an R-module, K is divisible, with R and M as submodules. It follows from Lemma 2.4 that the R-module K/M is divisible, and so injective. Thus R/M has an injective envelope E which is a submodule of K/M. Following the terminology of Proposition 4.23, we put $A_n = 0:_E M^n$ ($n = 1, 2, ...$). Then $MA_2 \subseteq A_1$, and $A_1 = R/M$ by Lemma 4.24. Since R/M is a simple R-module, either $MA_2 = 0$ or $MA_2 = A_1$.

Consider first the case $MA_2 = 0$. Then $A_2 \subseteq 0:_E M = A_1$, so $A_2 = A_1$. But Proposition 4.23 gives that $M = 0:A_1$ and $M^2 = 0:A_2$, so that $M = M^2 = M^3 = ...$ and so on. But, from Proposition 4.23 Corollary 1, $\bigcap_{n=1}^{\infty} M^n = 0$, so that $M = 0$. However, we are assuming that R is not a field.

We now know that $MA_2 = A_1$. Also,

$$K/M \supseteq E \supseteq A_2 \supseteq A_1 = R/M,$$

so there is an R-submodule T of K such that $T \supseteq R$ and $A_2 = T/M$. But $M(T/M) = R/M$, so that $(MT)/M = R/M$. (Note that $MT \supseteq MR = M$.) Thus $MT = R$, and we have found an inverse for M, since T will be a fractional ideal of R.

We must show that every non-zero integral ideal has an inverse. Consider the collection of all integral ideals of R which do not have inverses. This collection is not empty, having the zero ideal as a member. Since R is a Noetherian ring, the collection therefore possesses a maximal member I (say). Now I is contained in some maximal ideal M, and $I \subset M$, since we know that M has an inverse M^{-1} say. Further,

$$I = IM^{-1}M \subseteq IM^{-1} \subseteq MM^{-1} = R,$$

so that IM^{-1} is an integral ideal containing I. If $I \subset IM^{-1}$, then IM^{-1} has an inverse. But then $IM^{-1}(IM^{-1})^{-1} = R$, and I has inverse $M^{-1}(IM^{-1})^{-1}$. It follows that $I = IM^{-1}$, whence $IM = I$. Thus $I = IM = IM^2 = ...$, whence

$$I \subseteq \bigcap_{n=1}^{\infty} M^n = 0$$

by Proposition 4.23 Corollary 1. Thus every non-zero integral ideal of R has an inverse, as required. $\square$

4.4   Some further results

For the moment, we do not require the ring R to be commutative.

DEFINITION   *We say that R is a '(left) H-ring' if, whenever S_1 and S_2 are simple (left) R-modules such that*

$$\mathrm{Hom}_R\,(E(S_1),\, E(S_2)) \neq 0,$$

then $S_1 \approx S_2$.
We shall omit the word 'left' and refer simply to an H-ring.
Suppose that S_1 and S_2 are simple R-modules such that

$$\mathrm{Hom}_R\,(E(S_1),\, E(S_2)) \neq 0.$$

There exist maximal left ideals M_1 and M_2 of R such that $S_1 \approx R/M_1$ and $S_2 \approx R/M_2$, so that

$$\mathrm{Hom}_R\,(E(R/M_1),\, E(R/M_2)) \neq 0.$$

If R is commutative and Noetherian, then, because every maximal ideal is prime, Proposition 4.21 gives that $M_1 \subseteq M_2$, whence $M_1 = M_2$ and $S_1 \approx S_2$. Thus *every commutative Noetherian ring is an H-ring.*

Let A be a finitely embedded R-module. By Theorem 4.12, the zero submodule OA of A has a normal decomposition in A; and its associated indecomposable injectives in A are injective envelopes of simple modules. If R is an H-ring, then each of these associated indecomposable injectives is isolated. It follows from Theorem 4.13 that, if R is an H-ring, then OA has only one normal decomposition in A.

THEOREM 4.26   *Let R be an H-ring, let A be a finitely embedded R-module, and let $E(S_1), \ldots, E(S_n)$ be the reduced set of associated indecomposable injectives of OA in A, where the S_i are simple modules. Let*

$$OA = K_1 \cap \ldots \cap K_n$$

be the unique normal decomposition of OA in A, where K_i is $E(S_i)$-isotopic for $1 \leqslant i \leqslant n$. Then there exist unique submodules $A_1, \ldots, A_n$ of A such that

(i) $A = A_1 + \ldots + A_n$ (d.s.), *and*   (4.4.1)
(ii) OA_i *is an $E(S_i)$-isotopic submodule of A_i.*

Moreover, for $1 \leqslant i \leqslant n$,

$$A_i = \bigcap_{j \neq i} K_j \quad and \quad K_i = \sum_{j \neq i} A_j.$$

Proof. Put

$$A' = (A/K_1) \oplus \ldots \oplus (A/K_n)$$

and define the mapping $\alpha \colon A \to A'$ by

$$\alpha(a) = (a + K_1, \ldots, a + K_n),$$

where $a \in A$. Then α is a monomorphism. We shall show that α is an isomorphism. Put $B = A'/\alpha(A)$; we must show that $B = 0$. Now, if $B \neq 0$, then Proposition 2.24 shows that there is a simple R-module S and a non-zero homomorphism $B \to E(S)$. We shall show that, for every simple R-module S, the only homomorphism $B \to E(S)$ is the zero mapping.

Let S be a simple R-module and let $f \colon B \to E(S)$ be an R-homomorphism. Let $\eta \colon A' \to B$ be the natural mapping. To prove that $f = 0$, it suffices to show that $f\eta = 0$. Denote by

$$\phi_i \colon A/K_i \to A' \quad (1 \leqslant i \leqslant n)$$

the injection mappings. An arbitrary element of A' can be written as $\sum_{i=1}^{n} \phi_i(a_i + K_i)$, where the a_i belong to A. We must show that

$$\sum_{i=1}^{n} f\eta\phi_i(a_i + K_i) = 0.$$

The mapping $f\eta\phi_i \colon A/K_i \to E(S)$ can be extended to a homomorphism $\theta_i \colon E(A/K_i) \to E(S)$. But K_i is $E(S_i)$-isotopic, so

$E(A/K_i)$ is isomorphic to the direct sum of a finite number of copies of $E(S_i)$; let

$$\psi_i \colon E(S_i) \oplus \ldots \oplus E(S_i) \to E(A/K_i)$$

be such an isomorphism. Then we have a homomorphism

$$\theta_i \psi_i \colon E(S_i) \oplus \ldots \oplus E(S_i) \to E(S).$$

But R is an H-ring, so if $S_i \not\approx S$, the only homomorphism from $E(S_i)$ to $E(S)$ is the zero mapping. Thus, if $S_i \not\approx S$, $\theta_i \psi_i = 0$, whence also $\theta_i = 0$ and $f\eta\phi_i = 0$. Thus, if S is not isomorphic to any S_i, then $\sum\limits_{i=1}^{n} f\eta\phi_i(a_i + K_i) = 0$. On the other hand, if S is isomorphic, say, to S_1, then it cannot be isomorphic to $S_2, \ldots, S_n$, in which case

$$\sum_{i=1}^{n} f\eta\phi_i(a_i + K_i) = f\eta\phi_1(a_1 + K_1)$$
$$= \sum_{i=1}^{n} f\eta\phi_i(a_1 + K_i)$$
$$= f\eta\alpha(a_1)$$
$$= 0.$$

We have now established that α is an isomorphism. If we put $A_i = \alpha^{-1}\phi_i(A/K_i)$ for $1 \leqslant i \leqslant n$, then

$$A = A_1 + \ldots + A_n \quad \text{(d.s.)}$$

and $A_i \approx A/K_i$, so that OA_i is $E(S_i)$-isotopic.

To establish the uniqueness of the A_is, suppose that (4.4.1) holds, where OA_i is an $E(S_i)$-isotopic submodule of A_i, and consider the projections $\pi_i \colon A \to A_i$. Then

$$\bigcap_{i=1}^{n} \operatorname{Ker} \pi_i = OA. \qquad (4.4.2)$$

This intersection is irredundant because the A_i are not zero. Further, $A/\operatorname{Ker} \pi_i \approx A_i$, so that $\operatorname{Ker} \pi_i$ is $E(S_i)$-isotopic. Thus (4.4.2) is a normal decomposition of OA in A. By the uniqueness of such a decomposition, this means that $\operatorname{Ker} \pi_i = K_i$ for $1 \leqslant i \leqslant n$. Thus

$$A_i = \bigcap_{j \neq i} K_j,$$

which establishes the uniqueness of the expression (4.4.1). Also

$$K_i = \sum_{j \neq i} A_j,$$

and everything is proved. □

In the case of a commutative Noetherian ring, we can give an alternative description of the submodules A_i described in Theorem 4.26. We first prove an elementary lemma.

LEMMA 4.27 *Let R be a commutative ring, let M_1, M_2 be distinct maximal ideals of R and let k, l be positive integers. Then*

$$M_1^k + M_2^l = R.$$

Proof. If $M_1^k + M_2^l \neq R$, then there is a maximal ideal M such that $M_1^k + M_2^l \subseteq M$. But then $M_1 \subseteq M$ and $M_2 \subseteq M$, so that $M_1 = M = M_2$, which is not so. The result follows. □

We return to the situation of Theorem 4.26, except that now R is assumed to be a commutative Noetherian ring. Since A is finitely embedded, OA has associated prime ideals in A which are actually maximal ideals.

COROLLARY TO THEOREM 4.26 *Let R be a commutative Noetherian ring, let A be a finitely embedded R-module and let $M_1, ..., M_n$ be the associated prime ideals of OA in A. Then there exist unique submodules $A_1, ..., A_n$ of A such that*

$$A = A_1 + ... + A_n \quad (d.s.)$$

and such that OA_i is an M_i-isotopic submodule of A_i. Moreover, for $1 \leqslant i \leqslant n$,

$$A_i = \bigcup_{k=1}^{\infty} (0:_A M_i^k). \quad (4.4.3)$$

Proof. Comparing this with Theorem 4.26 itself, we see that it is only (4.4.3) that needs to be established. Now

$$A_i \subseteq E(A_i) \approx E(R/M_i) \oplus ... \oplus E(R/M_i),$$

where there are only finitely many terms in the direct sum. Since R is Noetherian, Proposition 4.23 shows that every element of $E(R/M_i)$ is annihilated by some positive power of M_i. The same can now be said of A_i, so that

$$A \subseteq_i \bigcup_{k=1}^{\infty} (0:_A M_i^k).$$

Conversely, consider an element $a \in A$ such that $M_i^k a = 0$. We can write $a = a_1 + \ldots + a_n$, where $a_j \in A_j$ for $1 \leqslant j \leqslant n$. Consider a fixed $j \neq i$. Then $M_i^k a_j = 0$ and, by the first part, $M_j^l a_j = 0$ for some positive integer l. It follows that $(M_i^k + M_j^l) a_j = 0$. But, by Lemma 4.27, $M_i^k + M_j^l = R$. Thus $a_j = 0$ for each $j \neq i$, and $a \in A_i$. This establishes (4.4.3).□

We can obtain from Theorem 4.26 an important decomposition theorem concerning commutative Artinian rings. But first we make some general remarks. Suppose that the ring R (not necessarily commutative) can be expressed in the form

$$R = R_1 + \ldots + R_n \quad \text{(d.s.)} \tag{4.4.4}$$

where $R_1, \ldots, R_n$ are two-sided ideals of R. Then 1_R can be written as

$$1_R = e_1 + \ldots + e_n,$$

where $e_k \in R_k$ for $1 \leqslant k \leqslant n$. If $r \in R_j$, then

$$re_1 + \ldots + re_n = r = e_1 r + \ldots + e_n r,$$

whence $re_j = r = e_j r$, since the sum (4.4.4) is direct. Thus e_j acts as an identity element for R_j and R_j has the structure of a ring with an identity element. Further, if $r_k \in R_k$ for $1 \leqslant k \leqslant n$ and $r \in R_j$, then

$$(r_1 + \ldots + r_n) r = r_j r$$

because $r_k r \in R_k \cap R_j$, which is the zero ideal when $k \neq j$. It follows that *the left ideals of R_j when R_j is regarded as a ring are the same as its submodules when it is regarded as a left ideal of R*. In this situation, we shall refer to R as the direct sum of the *rings* $R_1, \ldots, R_n$ rather than of the two-sided ideals $R_1, \ldots, R_n$.

THEOREM 4.28   *A commutative Artinian ring is a direct sum of a finite number of local Artinian rings.*

Proof. Let R be a commutative Artinian ring. Theorems 3.25 Corollary and 3.21 give that R is Noetherian and finitely embedded as an R-module. Denote by $M_1, \ldots, M_n (n \geqslant 0)$ the associated prime ideals of the zero ideal of R. These are actually maximal ideals of R. By Theorem 4.26 Corollary, there exist ideals $A_1, \ldots, A_n$ of R such that OA_i is an M_i-isotopic submodule of A_i $(1 \leqslant i \leqslant n)$ and

$$R = A_1 + \ldots + A_n \quad \text{(d.s.)}.$$

This expresses R as a direct sum of rings. Each A_i is an Artinian R-module and so also an Artinian ring.

It remains to show that each A_i is a local ring, or at least a quasi-local ring, since it is certainly Noetherian. Note that $A_i \neq 0$. Let M be a maximal ideal of A_i. Then M is a maximal submodule of A_i and A_i/M is a simple R-module. Consider Fig. 4.4 with obvious mappings, where $E(A_i)$ and $E(A_i/M)$ are injective envelopes of R-modules. This can be completed by an R-homomorphism ϕ as shown. But OA_i is an M_i-isotopic submodule of A_i, so that

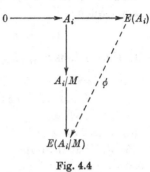

Fig. 4.4

$$E(A_i) \approx E(R/M_i) \oplus \ldots \oplus E(R/M_i),$$

where there are finitely many terms in the direct sum. Thus there is a non-zero homomorphism

$$E(R/M_i) \oplus \ldots \oplus E(R/M_i) \to E(A_i/M),$$

from which it follows that there is a non-zero homomorphism from $E(R/M_i)$ to $E(A_i/M)$. But R is an H-ring, so that

$$R/M_i \approx A_i/M.$$

Thus, if M and M' are both maximal ideals of A_i, then A_i/M and A_i/M' are isomorphic as R-modules, and so also as A_i-modules. It follows that $M = M'$, so that A_i has a unique maximal ideal. This is the ideal of non-units of A_i, and A_i is a quasi-local ring. $\square$

LEMMA 4.29   *Let R be a commutative ring, let E be an R-module and A a submodule of E. Let $r \in R$ and suppose that A and $0:_E r$ are both Artinian. Then $0:_{E/A} r$ is also Artinian.*

Proof. We define a mapping

$$\alpha: 0:_{E/A} r \to A/rA$$

by $\alpha(e + A) = re + rA$, where e is an element of E such that $re \in A$. Then α is well-defined and is an R-homomorphism. Let

$$e + A \in \operatorname{Ker} \alpha.$$

Then $re \in rA$, so that $e \in A + (0:_E r)$. Conversely, if $e \in A + (0:_E r)$, then $e + A \in \operatorname{Ker} \alpha$. Thus

$$\operatorname{Ker} \alpha = (A + (0:_E r))/A \approx (0:_E r)/((0:_E r) \cap A).$$

Thus $\operatorname{Ker} \alpha$ is Artinian; $\operatorname{Im} \alpha$ is also Artinian. It follows from the exact sequence

$$0 \to \operatorname{Ker} \alpha \to 0:_{E/A} r \to \operatorname{Im} \alpha \to 0,$$

with the obvious mappings, that $0:_{E/A} r$ is Artinian. $\square$

THEOREM 4.30   *Let R be a commutative Noetherian ring and let E be an R-module. Then the following statements are equivalent:*
   (a) *E is Artinian;*
   (b) *E is finitely embedded.*

Proof. That (a) implies (b) follows from Theorem 3.21, so we need to show that (b) implies (a); for this it is sufficient to prove that the R-module $E = E(R/M)$, where M is a maximal ideal of R, is Artinian. We suppose the contrary, and denote by Ω the collection of all ideals B of R such that $0:_E B$ is not Artinian. Then Ω is not empty, because $0:_E 0 = E$, so that the zero ideal is a member of Ω. Because R is Noetherian, Ω has a maximal member C (say). Now $(0:_E C) \cap (R/M) \neq 0$, so that

$$(0:_E C) \cap (R/M) = R/M,$$

because R/M is simple, and $0:_E C \supseteq R/M$. Thus $C \subseteq M$. But, by Lemma 4.24, $0:_E M = R/M$, which is certainly Artinian. Thus $C \subset M$.

Consider now the Noetherian ring R/C. It has a non-zero maximal ideal M/C. By Proposition 2.27, the (R/C)-module $0:_E C$ is an injective envelope of the (R/C)-module

$$(R/M) \cap (0:_E C) = R/M \approx (R/C)/(M/C).$$

Further, $0:_E C$ is not an Artinian (R/C)-module. Finally, if I is an ideal of R strictly containing C, then

$$0:_{0:_E C}(I/C) = 0:_E I$$

by Proposition 1.14, which is Artinian. If we now turn our attention from R, M and E to R/C, M/C and $0:_E C$, we see that we can impose the extra condition that every non-zero ideal of R has an Artinian annihilator in E.

Denote by Σ the set of all submodules A of E for which E/A is not finitely embedded. Then Σ is not empty, because E is not Artinian (Theorem 3.21). Partially order Σ by the opposite of inclusion, and let $\{A_\alpha\}_{\alpha \in \Lambda}$ be a totally ordered subset of Σ. Then $\{A_{\alpha'}/\bigcap_{\alpha \in \Lambda} A_\alpha\}_{\alpha' \in \Lambda}$ is an inverse system of submodules of $E/\bigcap_{\alpha \in \Lambda} A_\alpha$, and there is no non-zero submodule of $E/\bigcap_{\alpha \in \Lambda} A_\alpha$ which is contained in every $A_{\alpha'}/\bigcap_{\alpha \in \Lambda} A_\alpha$. In view of Proposition 3.19, there are two possibilities: either $A_{\alpha'}/\bigcap_{\alpha \in \Lambda} A_\alpha = 0$ for some $\alpha' \in \Lambda$ or else $E/\bigcap_{\alpha \in \Lambda} A_\alpha$ is not finitely embedded. In the former case,

$$\bigcap_{\alpha \in \Lambda} A_\alpha = A_{\alpha'} \quad \text{and} \quad E/\bigcap_{\alpha \in \Lambda} A_\alpha$$

is not finitely embedded, so both cases give the same conclusion, and the totally ordered subset of Σ is bounded above by $\bigcap_{\alpha \in \Lambda} A_\alpha$. Hence, by Zorn's Lemma, Σ has a maximal member A_0 (say). Note that A_0 is actually minimal with respect to inclusion.

Now A_0 is Artinian. To see this we use Theorem 3.21. Let A_1 be any submodule of A_0 and consider A_0/A_1. If $A_1 = A_0$, then A_0/A_1 is certainly finitely embedded. If $A_1 \subset A_0$, then, by the minimality of A_0, E/A_1 and hence also A_0/A_1 (Proposition 3.20) is finitely embedded. This shows that A_0 is Artinian. Let r be a non-zero element of M. In view of the extra assumption made in the course of the proof, $0:_E r$ is Artinian. It follows from Lemma 4.29 that $0:_{E/A_0} r$ is Artinian.

Denote by F an injective envelope of E/A_0. Since R is Noetherian, we can write

$$F = \sum_{i \in I} E_i \quad \text{(d.s.)}, \tag{4.4.5}$$

where the E_i are indecomposable injective modules (Theorem 4.4). Consider a particular $i \in I$. Now $E_i \cap (E/A_0) \neq 0$, so the composed mapping

$$E \to E/A_0 \xrightarrow{\ \text{inc}\ } F \xrightarrow{\ \text{proj}\ } E_i,$$

where $E \to E/A_0$ is the natural mapping, is not the zero mapping. Thus $\operatorname{Hom}_R(E, E_i) \neq 0$. But $E = E(R/M)$ and $E_i \approx E(R/P_i)$

for some prime ideal P_i of R (Theorem 2.32 Corollary). Thus
$$\mathrm{Hom}_R(E(R/M), E(R/P_i)) \neq 0,$$
and Proposition 4.21 shows that $M \subseteq P_i$. Thus $M = P_i$ and $E_i \approx E$. Now this is true for each i, and Propositions 3.16 and 3.15 give that
$$S(E/A_0) = S(F) \approx S(\underset{i \in I}{\oplus} E) = \underset{i \in I}{\oplus}(R/M). \qquad (4.4.6)$$
Thus $MS(E/A_0) = 0$. Since $r \in M$, this means that
$$S(E/A_0) \subseteq 0:_{E/A_0} r$$
and $S(E/A_0)$ is Artinian. Hence the direct sum in (4.4.6) must be a finite direct sum and the index set I is finite. Since $E_i \approx E(R/M)$, (4.4.5) gives that E/A_0 is finitely embedded. This is the contradiction for which we have been looking. $\square$

Exercises on Chapter 4

4.1 Show that every left module over a left Noetherian ring possesses a maximal injective submodule. Show also that every left module over a left Noetherian, left hereditary ring possesses an injective submodule which contains every injective submodule (see Exercise 2.11).

4.2 Let M be an R-module such that $E(M)$ is the direct sum of a finite number of its indecomposable injective submodules. Let N be a submodule of M. Show that $E(N)$ is the direct sum of a finite number of its indecomposable injective submodules. [*Hint:* Use Theorem 4.10]

4.3 Let M be an R-module such that $E(M)$ is an injective envelope of a direct sum of indecomposable injective R-modules and let N be a submodule of M. Show that $E(N)$ is an injective envelope of a direct sum of indecomposable injective R-modules. [*Hint:* Use Exercise 4.2]

4.4 Let E be an injective envelope of a direct sum of indecomposable injective R-modules and let F be an injective R-module which has no indecomposable injective submodule. If $\phi: E \to F$ is an R-homomorphism, show that E is an injective envelope of $\mathrm{Ker}\,\phi$. If $\psi: F \to E$ is an R-homomorphism, show that F is an injective envelope of $\mathrm{Ker}\,\psi$. [*Hint:* Use Exercise 4.3]

4.5 Let E be an injective R-module. Show that E can be expressed in the form $E = E_1 + E_2$ (d.s.), where E_1 is an injective envelope of a direct sum of indecomposable injective R-modules and E_2 has no indecomposable injective submodule. Show also that E_1 and E_2 are uniquely determined up to isomorphism. [*Hint:* For the uniqueness, suppose that

$$E_1 + E_2 \text{ (d.s.)} = E = E_1' + E_2' \text{ (d.s.)}$$

are two expressions of E in the given form. Consider the combined mappings

$$\alpha: E_1 \twoheadrightarrow E \twoheadrightarrow E_1',$$
$$\beta: E_1' \twoheadrightarrow E \twoheadrightarrow E_1,$$
$$\lambda: E_2 \twoheadrightarrow E \twoheadrightarrow E_2',$$
$$\mu: E_2' \twoheadrightarrow E \twoheadrightarrow E_2,$$

where each mapping is inclusion followed by projection. Use Exercise 4.3 to show that these mappings are monomorphisms. Use Exercise 4.4 to show that E_1 is an essential extension of $\beta\alpha(E_1)$ and that E_2 is an essential extension of $\mu\lambda(E_2)$.]

4.6 Let E be an injective R-module which is a direct sum of indecomposable injective submodules and let M be a direct summand of E. Show that M is also a direct sum of indecomposable injective submodules. [*Hint*: Use Exercises 4.3 and 3.5]

4.7 Let R be a commutative ring, let E be an R-module, let P be an ideal of R and N a submodule of E. Suppose that (a) $N \neq E$, (b) whenever $re \in N$, where $r \in R$ and $e \in E$, then either $e \in N$ or $r \in P$, (c) whenever $r \in P$, then $r^m E \subseteq N$ for some positive integer m. Show that P is a prime ideal of R and N is a P-primary submodule of E.

4.8 Let $n > 1$ be an integer. Describe the normal decomposition of the ideal Zn in the ring Z of integers. Why is there essentially only one normal decomposition of each ideal?

4.9 Let F be a field, let X, Y be independent indeterminates over F and let R denote the polynomial ring $F[X, Y]$. Show that

$$RX \cap (RX^2 + RY)$$

and
$$RX \cap (RX^2 + RXY + RY^2)$$

are normal decompositions of the ideal $RX^2 + RXY$. Comment on the fact that this ideal has more than one normal decomposition.

4.10 Let R be a commutative ring, M a maximal ideal of R and E a non-zero finitely generated R-module such that

$$E = \bigcup_{n=1}^{\infty} (0:_E M^n).$$

Show that the zero submodule of E is M-primary.

4.11 Let R be a commutative domain which is not a field and let M be a maximal ideal of R. Show that the zero submodule of $E(R/M)$ is M-isotopic but is not a primary submodule.

4.12 Let F be a field and let R be the ring of all polynomials in the independent indeterminates X_i $(i = 1, 2, 3, ...)$ with coefficients in F. Let I be the ideal of R generated by the elements $X_i X_j$ $(i, j = 1, 2, 3, ...)$, and let M be the ideal generated by the elements X_i $(i = 1, 2, 3, ...)$. Show that I is an M-primary ideal of R. Show also that M/I is the socle of the R-module R/I. Deduce that I is not an isotopic ideal of R.

4.13 Let R be a commutative ring, let P be a prime ideal of R and let $r \in R$. Show that the endomorphism ϕ of $E(R/P)$ given by $\phi(e) = re$, where $e \in E(R/P)$, is an isomorphism if and only if $r \notin P$.

4.14 Let R be a commutative ring, let P be a prime ideal of R and let $E = E(R/P)$. Put $A_n = 0:_E P^n$ for $n = 1, 2, 3, ...$.

(i) Show that the non-zero elements of A_{n+1}/A_n form the set of elements x of E/A_n such that $0: x = P$.

(ii) Let K be the quotient field of R/P. Show how A_{n+1}/A_n may be given the structure of a vector space over K and prove that $A_1 \approx K$ as vector spaces.

4.15 Let R be a commutative ring and let P be a prime ideal of R such that $E(R/P)$ is a Noetherian R-module. Show that P is a maximal ideal of R. [*Hint:* Use Proposition 2.27 and Exercise 1.1]

4.16 Let R be a commutative Noetherian ring. Show that R is Artinian if and only if every indecomposable injective R-module is Noetherian. [*Hint:* Use Exercise 4.15]

4.17 Let R be a commutative Noetherian ring and let A be an indecomposable injective R-module. Show that the following statements are equivalent: (*a*) A is Artinian; (*b*) $S(A) \neq 0$; (*c*) $A \approx E(R/M)$ for some maximal ideal M of R.

4.18 Let R be a commutative Noetherian ring and let A be an R-module. Show that A is Artinian if and only if there are maximal ideals $M_1, M_2, \ldots, M_n$ such that A can be embedded in $\overset{n}{\underset{i=1}{\oplus}} E(R/M_i)$.

4.19 Let A be a module over a commutative ring R. Prove that the following conditions are equivalent:

(*a*) A is finitely generated and Artinian;

(*b*) A is finitely embedded and Noetherian:

[*Hint:* Use Exercise 1.10]

4.20 Show that, over a commutative Artinian ring, a module is finitely embedded if and only if it is finitely generated.

Notes on Chapter 4

Theorem 4.1 is due to H. Bass and Z. Papp. In Theorem 4.4, the equivalence of (*a*) and (*b*) is due to E. Matlis and Z. Papp; the equivalence of (*a*) and (*c*) in a slightly weakened form is due to C. Faith and E. Walker. There are many interesting results about the decomposition of injective modules in C. Faith and E. Walker [6]. E. Matlis has posed the following question: if M is a direct summand of a direct sum of indecomposable injective modules, is M itself a direct sum of indecomposable injective modules? This question is still open; but see C. Faith and E. Walker [6] and also Exercise 4.6.

Most of the other results of this chapter are due to E. Matlis [19]. Using the ideas of E. Matlis, P. Gabriel [10] observed that indecomposable injective modules may be used to obtain various decomposition theorems, including the Lasker–Noether decomposition and the tertiary decomposition of L. Lesieur and R. Croisot [17] for modules over non-commutative rings.

Valuation rings form an interesting class of rings for which the decompositions are applicable (see E. Matlis [20]). In fact, the injective envelope of every finitely generated module over a valuation ring is the direct sum of a finite number of indecomposable injective modules. Note that another way of saying that $E(M)$ is the direct sum of n indecomposable injective modules is that M has so-called Goldie dimension n. (See C. Faith [5])

The characterization of Dedekind domains as those commutative domains R for which every divisible R-module is injective is due to H. Cartan and S. Eilenberg [3] (see Theorem 4.25). Their proof is more elementary than the one given here and involves the use of projective modules.

The characterization of Artinian modules over a commutative Noetherian ring as those modules which are finitely embedded is due to E. Matlis (Theorem 4.30). The following question arises: which commutative rings R have the property that the Artinian R-modules and the finitely embedded R-modules are one and the same? P. Vámos [24] has shown that these are the rings R such that the localization R_M is Noetherian for every maximal ideal M.

5. *Localization, completion and duality*

In this chapter we shall examine some connections between injective modules and localizations and completions of rings. We shall also establish a theory of duality for modules, over complete local rings, which satisfy either of the chain conditions.

5.1 Localization

We begin by describing briefly the classical construction of the localization of a commutative ring at a prime ideal. Suppose that we have a commutative ring R with a prime ideal P of R. We first introduce an equivalence relation $\sim$ on the set $R \times (R \backslash P)$;† we write $(r, t) \sim (r', t')$, where $r, r' \in R$ and $t, t' \in R \backslash P$, if these exists $t'' \in R \backslash P$ such that $t''t'r = t''tr'$. This equivalence relation partitions the set $R \times (R \backslash P)$; the set of equivalence classes is denoted by R_P; the equivalence class to which the element (r, t) of $R \times (R \backslash P)$ belongs is denoted by $[r, t]$. Addition and multiplication can be defined on R_P by

$$[r, t] + [r', t'] = [t'r + tr', tt'],$$

$$[r, t][r', t'] = [rr', tt'],$$

with an obvious notation. This gives R_P the structure of a commutative ring with zero element $[0, 1]$ and identity element $[1, 1]$. Note that $[0, 1] \neq [1, 1]$ because $(0, 1)$ and $(1, 1)$ are not equivalent.‡ The elements of R_P of the form $[r, t]$, where $r \in P$ and $t \in R \backslash P$, form a maximal ideal of R_P. Since this is the only maximal ideal of R_P, R_P is a quasi-local ring (see Section 3.2).

There is a mapping $\phi: R \to R_P$ given by $\phi(r) = [r, 1]$, where $r \in R$, and this mapping is a ring-homomorphism. We note three facts concerning ϕ:

† $R \backslash P$ stands for the set of all elements of R not in P and $\times$ denotes the Cartesian product.

‡ The ring R is non-trivial because it is assumed to possess the prime ideal P.

(i) Ker ϕ consists of all elements r of R for which there exists $t \in R \backslash P$ such that $tr = 0$;

(ii) $\phi(t)$ is a unit of R_P whenever $t \in R \backslash P$;

(iii) every element of R_P is of the form

$$[r, t] = [r, 1][t, 1]^{-1} = \phi(r)\phi(t)^{-1},$$

where $r \in R$ and $t \in R \backslash P$.

The construction whereby one forms R_P is termed *the localization of the commutative ring R with respect to the prime ideal P*.

We now make a completely fresh start. We shall show how the ring R_P may be defined in terms of injective modules. As a starting point, we shall use the properties (i), (ii) and (iii) above to define a localization of R with respect to P.

Throughout this section, we shall assume that we are given a commutative ring R and a prime ideal P of R. When we refer to a ring-homomorphism, we shall assume as part of the definition that it maps identity element to identity element. Also, when one ring is a subring of another, it is assumed that they have the same identity elements.

DEFINITION *A 'localization of R with respect to P' is a ring-homomorphism $\phi: R \to S$ satisfying the following conditions:*

(a) *Ker ϕ consists of all elements r of R for which there exists $t \in R \backslash P$ such that $tr = 0$;*

(b) *$\phi(t)$ is a unit of S whenever $t \in R \backslash P$;*

(c) *every element of S can be expressed in the form $\phi(r)\phi(t)^{-1}$, where $r \in R$ and $t \in R \backslash P$.*

We have not stated explicitly in the definition that the ring S is commutative; that it is can be seen in our first lemma.

LEMMA 5.1 *Let $\phi: R \to U$ be a ring-homomorphism, where U need not be a commutative ring, and suppose that $\phi(t)$ is a unit of U whenever $t \in R \backslash P$. Then the elements of U of the form $\phi(r)\phi(t)^{-1}$, where $r \in R$ and $t \in R \backslash P$, form a commutative subring of U.*

Proof. Let $r_1, r_2 \in R$. Then

$$\phi(r_1)\phi(r_2) = \phi(r_1 r_2) = \phi(r_2 r_1) = \phi(r_2)\phi(r_1).$$

Now let $t_1, t_2 \in R \backslash P$. Then

$$\phi(t_1)^{-1}\phi(t_2)^{-1} = (\phi(t_2)\phi(t_1))^{-1} = (\phi(t_1)\phi(t_2))^{-1} = \phi(t_2)^{-1}\phi(t_1)^{-1}.$$

Now let $r \in R$, $t \in R \backslash P$. Then

$$\phi(r)\,\phi(t)^{-1} = \phi(t)^{-1}\,\phi(t)\,\phi(r)\,\phi(t)^{-1} = \phi(t)^{-1}\,\phi(r)\,\phi(t)\,\phi(t)^{-1}$$
$$= \phi(t)^{-1}\,\phi(r).$$

It follows that any two elements of the form $\phi(r)\,\phi(t)^{-1}$, where $r \in R$ and $t \in R \backslash P$, commute. Further, with obvious notation,

$$\phi(r_1)\,\phi(t_1)^{-1} + \phi(r_2)\,\phi(t_2)^{-1} = \phi(r_1 t_2 + r_2 t_1)\,\phi(t_1 t_2)^{-1}, \quad (5.1.1)$$
$$-\phi(r)\,\phi(t)^{-1} = \phi(-r)\,\phi(t)^{-1},$$
$$(\phi(r_1)\,\phi(t_1)^{-1})\,(\phi(r_2)\,\phi(t_2)^{-1}) = \phi(r_1 r_2)\,\phi(t_1 t_2)^{-1} \qquad (5.1.2)$$

and $1 = \phi(1)\,\phi(1)^{-1}$, so that the elements $\phi(r)\,\phi(t)^{-1}$ form a commutative subring of U.□

If $\phi: R \to S$ is a localization of R with respect to P and if $\theta: S \to S'$ is a ring-isomorphism, then $\theta\phi: R \to S'$ is a localization of R with respect to P.

Now let $\phi: R \to S$ and $\phi': R \to S'$ be localizations of R with respect to P. A ring-homomorphism $\omega: S \to S'$ may be defined such that the diagram Fig. 5.1 is commutative. Indeed, there is only one possibility; we must put

Fig. 5.1

$$\omega(\phi(r)\,\phi(t)^{-1}) = \phi'(r)\,\phi'(t)^{-1},$$

where $r \in R$ and $t \in R \backslash P$. It may be verified that this does define a mapping from S to S' and that this mapping is a ring-isomorphism. This describes the extent to which localization of R with respect to P is unique.

We shall now establish the existence of a localization of R with respect to P. Put $E = E(R/P)$ and $H = \mathrm{Hom}_R(E, E)$, so that H is the ring of endomorphisms of E. We define the mapping $\phi: R \to H$ by

$$(\phi(r))\,(e) = re, \text{ where } r \in R \text{ and } e \in E.$$

Then ϕ is a ring-homomorphism. Note that

$$\mathrm{Ker}\,\phi = \mathrm{Ann}_R E.$$

LEMMA 5.2 *Let the situation be as just described and let t be an element of R. Then the following statements are equivalent;*

(a) $t \notin P$;

(b) $\phi(t)$ *is a unit of H*.

Proof. Assume that $t \notin P$. Now E is an indecomposable injective module so, by Lemma 3.10, (b) will follow if we show that $\operatorname{Ker} \phi(t) = 0$. Consider an element $x \in \operatorname{Ker} \phi(t) \cap (R/P)$. Then x is a coset $r + P$, where $r \in R$, and $tx = (\phi(t))(x) = 0$, whence $tr \in P$. But $t \notin P$, so that $r \in P$ and $x = 0$. Thus

$$\operatorname{Ker} \phi(t) \cap (R/P) = 0,$$

whence $\operatorname{Ker} \phi(t) = 0$, because E is an essential extension of R/P.

Now suppose that $t \in P$. Then, for $y \in R/P$, $(\phi(t))(y) = ty = 0$, so that $\operatorname{Ker} \phi(t) \supseteq R/P$. Thus $\operatorname{Ker} \phi(t) \neq 0$ and $\phi(t)$ is not a unit of H. $\square$

As a result of Lemma 5.2, we can now denote by S the set of all elements of H of the form $\phi(r) \phi(t)^{-1}$, where $r \in R$ and $t \in R \backslash P$. By Lemma 5.1, S is a commutative subring of H. Note that $\phi(R) \subseteq S$, so we may regard ϕ as a mapping from R into S.

PROPOSITION 5.3 *Let the situation be as just described. Then $\phi: R \to S$ is a localization of R with respect to P.*

Proof. Let $t \in R \backslash P$. Then $\phi(t)^{-1} = \phi(1) \phi(t)^{-1} \in S$, so that $\phi(t)$ is a unit of S. It remains to examine $\operatorname{Ker} \phi$. If $r \in \operatorname{Ker} \phi$, i.e. $r \in \operatorname{Ann}_R E(R/P)$, then Proposition 2.26 gives that there is an element $t \in R \backslash P$ such that $tr = 0$. Conversely, if $r \in R$ is such that $tr = 0$ for some $t \in R \backslash P$, then

$$\phi(r) = \phi(r) \phi(t) \phi(t)^{-1} = \phi(rt) \phi(t)^{-1} = 0$$

and $r \in \operatorname{Ker} \phi$. $\square$

We have now established the existence and essential uniqueness of a localization of R with respect to P. Strictly speaking, this is a ring-homomorphism $\phi: R \to S$. However, we shall allow the mapping itself to slide into the background and will refer to the ring S as a localization of R with respect to P; and we shall denote a localization of R with respect to P by R_P. Thus R_P will be a commutative ring. When we need to refer explicitly to the mapping, we shall call it the *canonical homomorphism* of R into R_P.

PROPOSITION 5.4 *R_P is a commutative quasi-local ring.*

Proof. We already know that the ring R_P is commutative, and may take as R_P the ring constructed in Proposition 5.3. It must be shown that the non-units of R_P form an ideal. Consider the element $\phi(r)\phi(t)^{-1}$ of R_P, where $r \in R$, $t \in R \backslash P$. If $r \notin P$, then $\phi(r)\phi(t)^{-1}$ has inverse $\phi(t)\phi(r)^{-1}$ in R_P. Conversely, suppose that $\phi(r)\phi(t)^{-1}$ is a unit of R_P. Then $\phi(r)$ is a unit of H and $r \notin P$ by Lemma 5.2. Thus the non-units of R_P are precisely those elements of the form $\phi(r)\phi(t)^{-1}$, where $r \in P$ and $t \in R \backslash P$. It follows from (5.1.1) and (5.1.2) that the non-units of R_P form an ideal.$\square$

NOTATION   We denote by PR_P the unique maximal ideal of R_P, so that PR_P consists of all elements of R_P of the form $\phi(r)\phi(t)^{-1}$, where $r \in P$ and $t \in R \backslash P$. The symbol PR_P really stands for $\phi(P)R_P$.

Let $\phi: R \to R_P$ be the canonical mapping and let A be an R_P-module. In a natural way, A has the structure of an R-module; we put
$$ra = \phi(r)\,a, \text{ where } r \in R \text{ and } a \in A.$$

In this way, every R_P-submodule of A is an R-submodule. Let B also be an R_P-module. Then an R_P-homomorphism $A \to B$ is also an R-homomorphism. Conversely consider an R-homomorphism $\alpha: A \to B$. For every $t \in R \backslash P$, $a \in A$, we have

$$\phi(t)\,\alpha(\phi(t)^{-1}a) = t\alpha(\phi(t)^{-1}a) = \alpha(t\phi(t)^{-1}a)$$
$$= \alpha(\phi(t)\,\phi(t)^{-1}a)$$
$$= \alpha(a),$$

so that
$$\alpha(\phi(t)^{-1}a) = \phi(t)^{-1}\alpha(a).$$

It follows from this that α is also an R_P-homomorphism. Thus *the R-homomorphisms $A \to B$ and the R_P-homomorphisms are the same,* i.e.
$$\operatorname{Hom}_R(A, B) = \operatorname{Hom}_{R_P}(A, B).$$

Now $E = E(R/P)$ is an R-module. Since R_P may be taken as a subring of the ring of endomorphisms of E, we may give E the structure of an R_P-module; we put
$$\rho e = \rho(e), \text{ where } \rho \in R_P \text{ and } e \in E.$$

Note that, if $r \in R$ and $e \in E$, then
$$\phi(r)\,e = (\phi(r))\,(e) = re,$$

so that, if having regarded E as an R_P-module we convert it into an R-module by using the homomorphism $\phi \colon R \to R_P$ as explained above, we merely recover the R-module structure of E that we had before.

PROPOSITION 5.5   *Let $\phi \colon R \to R_P$ be a localization of R with respect to P and let A be an R_P-module. Then the following statements are equivalent:*

    (a) *A is injective as an R_P-module;*

    (b) *A is injective as an R-module.*

Proof. Suppose first that A is injective as an R-module and let B be an R_P-module which extends A. To prove that A is injective as an R_P-module, it is sufficient to show that A is a direct summand of B (Theorem 2.15). Certainly, if we regard A, B as R-modules, then A is a direct summand of B, so that we can write $B = A + C$ (d.s.), where C is an R-submodule of B. We must show that C is an R_P-submodule of B. Let $t \in R \backslash P$ and let $c \in C$. Then $\phi(t)^{-1} c = a + c'$ for some $a \in A$, $c' \in C$, whence $c = \phi(t) a + \phi(t) c'$. But $\phi(t) c' = tc' \in C$ and the sum $A + C$ is direct, so that $c = \phi(t) c'$ and $\phi(t)^{-1} c = c' \in C$. Thus, if $r \in R$, then $\phi(r) \phi(t)^{-1} c \in C$ and C is an R_P-submodule of B.

Before we prove the converse implication, we need to consider what we have just proved as applied to $E(R/P) = E$. We can consider E as an R_P-module; as an R-module it is injective, so it is injective as an R_P-module. Now E is indecomposable as an R-module and its R_P-submodules are also R-submodules, so that E is also indecomposable as an R_P-module. Let x be a non-zero element of R/P. Then x is annihilated by every non-unit of R_P, and the annihilator of x in R_P is just PR_P, the unique maximal ideal of R_P. Thus $R_P x \approx R_P/PR_P$. Now E is an injective envelope of $R_P x$ as an R_P-module. We shall use the symbol $\bar{E}(R_P/PR_P)$ to denote an injective envelope of R_P/PR_P as an R_P-module. Thus we have
$$E(R/P) \approx \bar{E}(R_P/PR_P),$$
the isomorphism being an R_P-isomorphism.

We state this as a separate result.

PROPOSITION 5.6   *As an R_P-module, $E(R/P)$ is isomorphic to an injective envelope of R_P/PR_P, where PR_P denotes the unique maximal ideal of R_P.* $\square$

We can now complete the proof of Proposition 5.5. Suppose that A is injective as an R_P-module. Every simple R_P-module is isomorphic to R_P/PR_P, and Proposition 2.25 Corollary gives that A can be embedded in a direct product T of copies of $\bar{\bar{E}}(R_P/PR_P)$. Then A is a direct summand of T; this is also so when A and T are regarded as R-modules. But

$$E(R/P) \approx \bar{\bar{E}}(R_P/PR_P)$$

as R_P-modules, so as an R-module T is isomorphic to a direct product of copies of $E(R/P)$ and so is injective. It follows that A is injective as an R-module.$\square$

COROLLARY TO PROPOSITION 5.5 *If R is a Noetherian ring, then so is R_P.*

Proof. Let R be Noetherian. We use the characterization of Noetherian rings given in Theorem 4.1. Let $\{E_i\}_{i \in I}$ be a family of injective R_P-modules. Then each E_i is an injective R-module, so that $\underset{i \in I}{\oplus} E_i$ is an injective R-module. It follows that $\underset{i \in I}{\oplus} E_i$ is also an injective R_P-module.$\square$

5.2 The ring of endomorphisms

In this section we shall discuss the ring of endomorphisms of an indecomposable injective module over a commutative Noetherian ring. This is preparatory to the section on the completion of a local ring.

We first prove four very general 'diagram-chasing' lemmas which could equally well have been dealt with in Chapter 1. The ring need not be commutative for these.

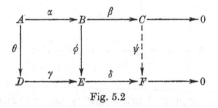

Fig. 5.2

LEMMA 5.7 *Given the commutative diagram Fig. 5.2 of R-modules and R-homomorphisms, with exact rows, there is a unique mapping $\psi: C \to F$ which makes the resulting diagram* *commutative, and ψ is an R-homomorphism. Further, if θ is an epimorphism and ϕ is an isomorphism, then ψ is an isomorphism.*

Proof. The mapping ψ must satisfy the condition $\psi\beta = \delta\phi$. Since β is an epimorphism, this specifies ψ uniquely. To see that there is such a mapping, suppose that b_1, b_2 of B are such that $\beta(b_1) = \beta(b_2)$. Then

$$b_1 - b_2 \in \operatorname{Ker}\beta = \operatorname{Im}\alpha,$$

so there exists $a \in A$ such that $\alpha(a) = b_1 - b_2$. Then

$$\delta\phi(b_1 - b_2) = \delta\phi\alpha(a) = \delta\gamma\theta(a) = 0$$

since $\delta\gamma$ is a zero mapping. Thus $\delta\phi(b_1) = \delta\phi(b_2)$ and ψ is well-defined. We leave the reader to verify that ψ is an R-homomorphism.

Now suppose that θ is an epimorphism and ϕ an isomorphism. Suppose that $\psi\beta(b) = 0$ for some $b \in B$. Then $\delta\phi(b) = 0$, so that $\phi(b) = \gamma(d)$ for some $d \in D$. There exists $a' \in A$ such that $\theta(a') = d$. Then
$$\phi(b) = \gamma\theta(a') = \phi\alpha(a').$$
Hence $\qquad b = \alpha(a') \quad \text{and} \quad \beta(b) = \beta\alpha(a') = 0.$

This shows that ψ is a monomorphism. Since δ and ϕ are both epimorphisms, it is easy to see that ψ is an epimorphism.□

LEMMA 5.8 *Given the commutative diagram Fig. 5.3 of R-modules and R-homomorphisms, with exact rows, there is a unique mapping $\theta: A \to D$ which makes the resulting diagram commutative, and θ is an R-homomorphism. Further, if ϕ is a monomorphism, then θ is a monomorphism.*

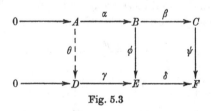

Fig. 5.3

Proof. The mapping θ must satisfy the condition $\gamma\theta = \phi\alpha$. Since γ is a monomorphism, this specifies θ uniquely. To see that there is such a mapping, let $a \in A$. Then

$$\delta\phi\alpha(a) = \psi\beta\alpha(a) = 0,$$

so that $$\phi\alpha(a)\in \mathrm{Ker}\,\delta = \mathrm{Im}\,\gamma$$

and there exists a unique element $d \in D$ such that $\gamma(d) = \phi\alpha(a)$. We put $\theta(a) = d$. We leave the reader to verify that θ is an R-homomorphism. The last part follows easily from the fact that $\phi\alpha = \gamma\theta.\square$

LEMMA 5.9 *In the commutative diagram Fig.* 5.4 *of R-modules and R-homomorphisms, suppose that* $\mathrm{Ker}\,\delta \subseteq \mathrm{Im}\,\gamma$ *and that* β, θ, ψ *are epimorphisms. Then* ϕ *is an epimorphism.*

Proof. Let $e \in E$. There exists $b \in B$ such that $\psi\beta(b) = \delta(e)$. Then

$$\delta\phi(b) = \delta(e),$$

so that $$e - \phi(b) \in \mathrm{Ker}\,\delta \subseteq \mathrm{Im}\,\gamma.$$

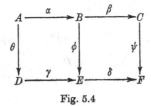

Fig. 5.4

There now exists $d \in D$ such that $\gamma(d) = e - \phi(b)$. In turn there exists $a \in A$ such that $\theta(a) = d$. Now

$$\phi(b + \alpha(a)) = e - \gamma(d) + \phi\alpha(a) = e - \gamma(d) + \gamma\theta(a) = e.$$

This shows that ϕ is an epimorphism.$\square$

LEMMA 5.10 *In the commutative diagram Fig.* 5.4 *of R-modules and R-homomorphisms, suppose that* ϕ *is an epimorphism,* ψ *and* γ *are monomorphisms,* $\delta\gamma = 0$ *and* $\mathrm{Ker}\,\beta \subseteq \mathrm{Im}\,\alpha$. *Then* θ *is an epimorphism.*

Proof. Let $d \in D$. There exists $b \in B$ such that $\phi(b) = \gamma(d)$, whence
$$\psi\beta(b) = \delta\phi(b) = \delta\gamma(d) = 0.$$

Hence $\beta(b) = 0$, so there exists $a \in A$ such that $\alpha(a) = b$. Then

$$\gamma\theta(a) = \phi\alpha(a) = \phi(b) = \gamma(d),$$

so that $\theta(a) = d$ and θ is an epimorphism.$\square$

We next recall the concept of the dual of a vector space. Let V be a vector space over the field K. Then $\mathrm{Hom}_K(V, K)$ is called the *dual space* of V. Suppose that V has a finite basis $v_1, v_2, \ldots, v_r$ (say). For $1 \leqslant i \leqslant r$, we define $v_i^* \in \mathrm{Hom}_K(V, K)$ by $v_i^*(v_j) = \delta_{ij}$, i.e. $v_i^*(v_j) = 0$ if $i \neq j$ and $v_i^*(v_i) = 1$. Then $v_1^*, v_2^*, \ldots, v_r^*$ is a basis for the dual space of V. Thus, *if V is a finitely generated vector space, then it is isomorphic to its dual space.*

Now let R be a commutative Noetherian ring and let E be an indecomposable injective R-module. By Theorem 2.32 Corollary, $E \approx E(R/P)$ for some prime ideal P of R. We shall consider in this section the ring $\mathrm{Hom}_R(E, E)$ of endomorphisms of E. If we localize the ring R with respect to P, we obtain the local ring R_P, and $E(R/P)$ can be regarded as an R_P-module. Further, the endomorphisms of $E(R/P)$ as an R_P-module are the same as its endomorphisms as an R-module, i.e.

$$\mathrm{Hom}_R(E(R/P), E(R/P)) = \mathrm{Hom}_{R_P}(E(R/P), E(R/P)).$$

Finally, as an R_P-module, $E(R/P) \approx \bar{E}(R_P/PR_P)$. All this information comes from the previous section. Thus there is a ring-isomorphism

$$\mathrm{Hom}_R(E, E) \approx \mathrm{Hom}_{R_P}(\bar{E}(R_P/PR_P), \bar{E}(R_P/PR_P)).$$

Because of this, *we can impose the extra conditions that R is a local ring and that $E = E(R/M)$, where M is the maximal ideal of R.* These conditions will be assumed for the time being. We shall make a temporary abbreviation of terminology. When A and B are R-modules, we shall denote the R-module $\mathrm{Hom}_R(A, B)$ by $H(A, B)$.

We consider the chains

$$R = M^0 \supseteq M \supseteq M^2 \supseteq \ldots,$$

$$0 = A_0 \subseteq A_1 \subseteq A_2 \subseteq \ldots,$$

where $A_n = 0:_E M^n$ for $n = 0, 1, 2, \ldots$. For each $n \geqslant 1$ we have the exact sequences

$$0 \to M^{n-1}/M^n \to R/M^n \to R/M^{n-1} \to 0,$$

$$0 \to A_{n-1} \to A_n \to A_n/A_{n-1} \to 0,$$

with the obvious mappings, and the former gives rise to the exact sequence

$$0 \to H(R/M^{n-1}, E) \to H(R/M^n, E) \to H(M^{n-1}/M^n, E) \to 0$$

because E is injective. For $n = 0, 1, 2, \ldots$, we define the mapping

$$\phi_n: H(R/M^n, E) \to A_n$$

by $\phi_n(\alpha) = \alpha(1 + M^n)$, where $\alpha \in H(R/M^n, E)$. It is easily checked that ϕ_n is an R-isomorphism. Further, Fig. 5.5 is commutative

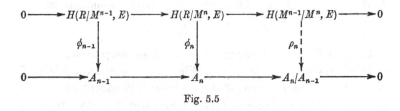

Fig. 5.5

for $n \geqslant 1$. It follows from Lemma 5.7 that there is a unique mapping
$$\rho_n: H(M^{n-1}/M^n, E) \to A_n/A_{n-1}$$

which makes the resulting diagram commutative, and that ρ_n is an R-isomorphism. Now

$$MH(M^{n-1}/M^n, E) = 0 = M(A_n/A_{n-1}),$$

so that both $H(M^{n-1}/M^n, E)$ and A_n/A_{n-1} may be regarded as modules over the field R/M, i.e. as vector spaces over R/M. Then ρ_n is also an (R/M)-isomorphism. If $\alpha \in H(M^{n-1}/M^n, E)$, then $\text{Im } \alpha \subseteq A_1$, from which it follows that the two vector spaces $H(M^{n-1}/M^n, E)$ and $H(M^{n-1}/M^n, A_1)$ are isomorphic. [Let $\beta \in H(M^{n-1}/M^n, A_1)$ correspond to $i\beta \in H(M^{n-1}/M^n, E)$, where $i: A_1 \to E$ is the inclusion mapping.] By Lemma 4.24, $A_1 = R/M$. Thus the vector space A_n/A_{n-1} is isomorphic to the dual space of M^{n-1}/M^n. But R is a Noetherian ring so that M^{n-1}/M^n is finitely generated. It follows that A_n/A_{n-1} is isomorphic to M^{n-1}/M^n. In turn $H(A_n/A_{n-1}, E)$ is isomorphic to the dual space of A_n/A_{n-1}, so that *$H(A_n/A_{n-1}, E)$ and M^{n-1}/M^n are isomorphic finite-dimensional vector spaces over R/M.* This fact will be useful later.

For each $n \geqslant 0$, we can define an R-homomorphism

$$\tau_n \colon R \to H(A_n, A_n)$$

by $(\tau_n(r))(a) = ra$, where $r \in R$ and $a \in A_n$, and

$$\operatorname{Ker} \tau_n = 0 \colon A_n = M^n$$

by Proposition 4.23. Thus τ_n induces an R-monomorphism

$$R/M^n \to H(A_n, A_n).$$

Now if $\alpha \in H(A_n, E)$, then $\operatorname{Im} \alpha \subseteq A_n$, from which it follows that $H(A_n, A_n) \approx H(A_n, E)$. Thus we have an R-monomorphism

$$\psi_n \colon R/M^n \to H(A_n, E).$$

LEMMA 5.11   *The mappings ψ_n are R-isomorphisms.*

Proof. We use induction on n. When $n = 0$ the result is trivial. Now let $n > 0$ and assume that ψ_{n-1} is an R-isomorphism. There is a commutative diagram, Fig. 5.6, and this has exact rows. It follows from Lemma 5.8 that there is a mapping

$$\chi_n \colon M^{n-1}/M^n \to H(A_n/A_{n-1}, E)$$

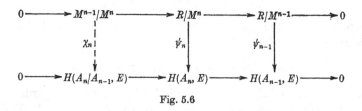

Fig. 5.6

which makes the resulting diagram commutative, and that χ_n is an R-monomorphism. But χ_n is also an (R/M)-monomorphism, and we know that the (R/M)-spaces M^{n-1}/M^n and $H(A_n/A_{n-1}, E)$ have the same finite dimension. It follows that χ_n is an isomorphism. But now Lemma 5.9 gives that ψ_n is an epimorphism, and so also an isomorphism. $\square$

PROPOSITION 5.12   *Let R be a local ring with maximal ideal M and let $E = E(R/M)$. Let ϕ be an endomorphism of E and let n be a positive integer. Then there exists $r_n \in R$ such that $\phi(e) = r_n e$ for all $e \in A_n$, where $A_n = 0 \colon_E M^n$.*

Proof. Denote by ϕ_n the restriction of ϕ to A_n, so that $\phi_n \in H(A_n, E)$. By Lemma 5.11, there exists $r_n \in R$ such that $\psi_n(r_n + M^n) = \phi_n$. This means that, for $e \in A_n$,

$$\phi(e) = \phi_n(e) = r_n e. \; \square$$

COROLLARY *Let R be a local ring with maximal ideal M and let $E = E(R/M)$. Let ϕ be an endomorphism of E and let $e \in E$. Then there exists an element r of R (dependent upon e) such that $\phi(e) = re$.*

Proof. This follows because $E = \bigcup\limits_{n=1}^{\infty} A_n$ (see Proposition 4.23). $\square$

PROPOSITION 5.13 *Let R be a commutative Noetherian ring and let E be an indecomposable injective R-module. Then the ring of endomorphisms of E is a commutative ring.*

Proof. By our earlier remarks, we may suppose that R is a local ring with maximal ideal M and that $E = E(R/M)$. Let ϕ, ψ be endomorphisms of E and let $e \in E$. Then, by Proposition 5.12 Corollary, there exist $r, s \in R$ such that $\phi(e) = re$, $\psi(e) = se$. Then

$$(\phi\psi)(e) = \phi(se) = s\phi(e) = s(re) = (sr)e = (rs)e = (\psi\phi)(e),$$

so that $\phi\psi = \psi\phi$. $\square$

5.3 Completion

We begin this section with some general remarks about a special class of topological rings. We assume that the reader is familiar with the basic notions of topology, although these will be recalled implicitly as we consider the specialized situation.

Let R be a ring (not necessarily commutative) and let

$$I_1 \supseteq I_2 \supseteq I_3 \supseteq \dots$$

be a descending chain of two-sided ideals of R. Such a chain is called a *filtration* of R and can be used to give R the structure of a topological space. A subset U of R is said to be *open* if, whenever $r \in U$, then there exists n such that $r + I_n \subseteq U$. The empty set and R itself are open, the union of an arbitrary family of open subsets of R is open and the intersection of a finite number of open subsets is open. These are precisely the requirements for R to be a *topological space*. We note also that the sets $r + I_n$, where $r \in R$, are open subsets of R. In particular, the two-sided ideals

$$I_n \quad (n = 1, 2, \ldots)$$

are open in R.

We endow the Cartesian product $R \times R$ with the product topology. Thus an open subset of $R \times R$ is the union of a family of sets of the form $U_1 \times U_2$, where U_1 and U_2 are open subsets of R. The axioms for a topological space are easily verified. We now define the mappings

$$\alpha, \beta \colon R \times R \to R \quad \text{and} \quad \gamma \colon R \to R$$

by
$$\alpha(r_1, r_2) = r_1 + r_2, \quad \beta(r_1, r_2) = r_1 r_2,$$
$$\gamma(r) = -r,$$

where $r_1, r_2, r \in R$. We assert that α, β, γ are continuous mappings. To show that α is continuous, we consider $r_1, r_2 \in R$ and an open subset V of R containing $r_1 + r_2$. We must find an open subset U of $R \times R$ containing (r_1, r_2) such that $\alpha(U) \subseteq V$. Since V is open, there exists n such that $r_1 + r_2 + I_n \subseteq V$. But

$$\alpha(r_1 + I_n, r_2 + I_n) \subseteq (r_1 + r_2) + I_n \subseteq V,$$

so we can take $U = (r_1 + I_n) \times (r_2 + I_n)$. It follows in a similar way from the inclusions

$$\beta(r_1 + I_n, r_2 + I_n) \subseteq (r_1 r_2) + I_n,$$
$$\gamma(r + I_n) \subseteq (-r) + I_n,$$

where $r_1, r_2, r \in R$, that β and γ are also continuous. Thus the ring operations of R are continuous and R is thereby a *topological ring*.

Suppose that $\bigcap_{n=1}^{\infty} I_n = 0$ and let $r, s \in R$, $r \neq s$. Then there exists n such that $r - s \notin I_n$. It follows that $r + I_n$ and $s + I_n$ do not meet. We have thus found disjoint open subsets U and V of R such that $r \in U$, $s \in V$. This says that R is a Hausdorff space. On the other hand, if $\bigcap_{n=1}^{\infty} I_n \neq 0$, then there exists $r \in \bigcap_{n=1}^{\infty} I_n$, $r \neq 0$, and every open set which contains 0 also contains r. Thus *R is a Hausdorff space if and only if $\bigcap_{n=1}^{\infty} I_n = 0$.*

A sequence $\{r_n\}$ of elements of R is *convergent with limit* r if, given a positive integer k, there exists an integer N such that $r_n - r \in I_k$ whenever $n \geqslant N$. Suppose that the sequence $\{r_n\}$ has limits r and s in R. Then, given k, there exists N such that $r_N - r \in I_k, r_N - s \in I_k$, so that $r - s \in I_k$. This is true for all k, so that $r - s \in \bigcap_{k=1}^{\infty} I_k$. Thus, *if R is Hausdorff, a sequence has at most one limit.* If the sequence $\{r_n\}$ has limit r and $\{s_n\}$ has limit s, then, as in elementary analysis, the sequence $\{r_n + s_n\}$ has limit $r + s$ and $\{r_n s_n\}$ has limit rs.

A sequence $\{r_n\}$ of elements of R is a *Cauchy sequence* if, given a positive integer k, there exists an integer N such that

$$r_m - r_n \in I_k$$

whenever $m, n \geqslant N$. Every convergent sequence is a Cauchy sequence. If every Cauchy sequence is convergent and if, further, R is Hausdorff, then R is said to be *complete* in its topology.

An oft-occurring example of a topological ring whose topology arises from a filtration is provided by a local ring with filtration the powers of its maximal ideal. This topology is called the *natural topology* of the local ring, and is Hausdorff by Proposition 4.23 Corollary 2. If we say that a local ring is 'complete', we shall mean that it is complete in its natural topology.

A *closed* subset of R is just the complement of an open subset. Thus the empty set and R itself are closed, the intersection of an arbitrary family of closed subsets is closed and the union of a finite number of closed subsets is closed. Further,

$$R \backslash I_n = \bigcup (r + I_n),$$

where the union is over all $r \in R \backslash I_n$. Thus $R \backslash I_n$ is open, so that I_n is closed. Thus *the two-sided ideals I_n ($n = 1, 2, 3, \ldots$) are both open and closed subsets of R.*

Let X be a subset of R. The *closure* of X, which is denoted by $\bar{X}$, is the intersection of all the closed subsets of R which contain X. Then $\bar{X}$ is a closed subset of R. Let $r \in \bar{X}$. Then, given any positive integer n, $r + I_n$ must meet X, for otherwise the complement of $r + I_n$ is a closed subset of R which contains X yet does not contain r. Conversely, suppose that $r \in R$ is such that $r + I_n$ meets

X for every n and let C be a closed subset of R containing X. Then $r \in C$, for otherwise $r \in R \setminus C$, which is open in R, and there exists n such that $r + I_n \subseteq R \setminus C$. But then $r + I_n$ would not meet X. It follows that $r \in \overline{X}$. Hence $\overline{X}$ consists of all $r \in R$ for which $r + I_n$ meets X for every n. Equivalently, $\overline{X}$ *consists of all elements of R which are limits of sequences of elements of X.* If $\overline{X} = R$, then X is said to be *dense* in R.

Let R, R' be rings with respective filtrations

$$I_1 \supseteq I_2 \supseteq I_3 \supseteq \dots \quad \text{and} \quad I_1' \supseteq I_2' \supseteq I_3' \supseteq \dots.$$

Let $\omega \colon R \to R'$ be a continuous mapping (not necessarily a ring-homomorphism) and let $\{r_n\}$ be a sequence of elements of R with limit r. Let the positive integer k be given. Then there exists l such that

$$\omega(r + I_l) \subseteq \omega(r) + I_k'.$$

Now $r_n - r \in I_l$ for all sufficiently large n, so that

$$\omega(r_n) \in \omega(r) + I_k'$$

for all sufficiently large n. Thus *the sequence* $\{\omega(r_n)\}$ *is convergent with limit* $\omega(r)$. This is of course true in the context of general topological spaces.

Let R, $\hat{R}$ be rings with respective filtrations

$$I_1 \supseteq I_2 \supseteq I_3 \supseteq \dots \quad \text{and} \quad \hat{I}_1 \supseteq \hat{I}_2 \supseteq \hat{I}_3 \supseteq \dots.$$

These filtrations give R and $\hat{R}$ the structures of topological rings. Let $\phi \colon R \to \hat{R}$ be a ring-homomorphism.

DEFINITION *The pair* $(\hat{R}, \phi)$ *is said to be a 'completion' of R if:*

   (a) $\hat{R}$ *is complete;*

   (b) $\operatorname{Ker} \phi = \bigcap\limits_{n=1}^{\infty} I_n$;

   (c) $\phi(R)$ *is dense in* $\hat{R}$;

   (d) $\phi(I_n) = \phi(R) \cap \hat{I}_n$ *for all* n.

Note that, when $(\hat{R}, \phi)$ is a completion of R, then $\phi(I_n) \subseteq \hat{I}_n$, whence $\phi(r + I_n) \subseteq \phi(r) + \hat{I}_n$ for all $r \in R$. Thus ϕ is a continuous mapping. Also, if R itself is complete, then (R, id) is a completion of R.

Let $(\hat{R}, \phi)$ be a completion of R. Since $\hat{I}_n$ is closed in $\hat{R}$, the closure $\overline{\phi(I_n)}$ of $\phi(I_n)$ in $\hat{R}$ is contained in $\hat{I}_n$. Conversely, consider $\hat{r} \in \hat{I}_n$. Because $\phi(R)$ is dense in $\hat{R}$, we can write $\hat{r} = \lim \phi(r_m)$ for some sequence $\{r_m\}$ in R, and there exists N such that $\phi(r_m) \in \hat{I}_n$ whenever $m \geqslant N$. Thus $\phi(r_m) \in \phi(R) \cap \hat{I}_n = \phi(I_n)$ for $m \geqslant N$, whence $r_m \in I_n$ for $m \geqslant N$ since $I_n \supseteq \mathrm{Ker}\,\phi$. This means that $\hat{r}$ is the limit of a sequence of elements of $\phi(I_n)$, so that $\hat{I}_n \subseteq \overline{\phi(I_n)}$. Thus $\hat{I}_n$ *is the closure in $\hat{R}$ of $\phi(I_n)$.*

We still suppose that $(\hat{R}, \phi)$ is a completion of R. Suppose further that R is a commutative ring. Consider $\hat{r}, \hat{s} \in \hat{R}$. We can write $\hat{r} = \lim \phi(r_n)$, $\hat{s} = \lim \phi(s_n)$ for some sequences $\{r_n\}$, $\{s_n\}$ of elements of R. Then

$$\hat{r}\hat{s} = \lim (\phi(r_n)\,\phi(s_n)) = \lim \phi(r_n s_n) = \lim \phi(s_n r_n) = \hat{s}\hat{r}.$$

Thus $\hat{R}$ is commutative. Disregarding ϕ, we say that *the completion of a commutative ring is again a commutative ring.*

We now consider the extent to which the completion of a ring is unique. Let $\hat{R}'$ be another ring with filtration $\{\hat{I}_n'\}$, and let $\phi': R \to \hat{R}'$ be a ring-homomorphism. Suppose that $(\hat{R}', \phi')$ as well as $(\hat{R}, \phi)$ is a completion of R. We shall construct a continuous mapping $\omega: \hat{R} \to \hat{R}'$ which makes Fig. 5.7 commutative. Indeed, there is only one possibility for ω. For let ω be such a mapping and consider $\hat{r} \in \hat{R}$. There exists a sequence $\{r_n\}$ of elements of R such that $\lim \phi(r_n) = \hat{r}$. Then the sequence

Fig. 5.7

$$\{\omega\phi(r_n)\} = \{\phi'(r_n)\}$$

is convergent with limit $\omega(\hat{r})$. Thus we must have

$$\omega(\hat{r}) = \lim \phi'(r_n). \qquad (5.3.1)$$

We must now show that (5.3.1) does define a mapping from $\hat{R}$ to $\hat{R}'$. Let $\{r_n\}$ be a sequence of elements of R such that $\{\phi(r_n)\}$ is a convergent sequence of $\hat{R}$ with limit $\hat{r}$. Then $\{\phi(r_n)\}$ is a Cauchy sequence, i.e. given a positive integer k, there exists N such that

$$\phi(r_m - r_n) = \phi(r_m) - \phi(r_n) \in \hat{I}_k \quad \text{for all} \quad m, n \geqslant N.$$

Thus, for $m, n \geqslant N$,

$$\phi(r_m - r_n) \in \phi(R) \cap \hat{I}_k = \phi(I_k),$$

so that $r_m - r_n \in I_k$ and $\{r_n\}$ is a Cauchy sequence of R. This in turn means that $\{\phi'(r_n)\}$ is a Cauchy sequence in $\hat{R}'$ and so is convergent. We put

$$\omega(\hat{r}) = \lim \phi'(r_n).$$

This is still insufficient to guarantee the existence of ω; we need to see what happens when we take another sequence $\{r_n'\}$ of elements of R such that $\lim \phi(r_n') = \hat{r}$. Then

$$\lim \phi(r_n' - r_n) = \lim \phi(r_n') - \lim \phi(r_n) = 0,$$

from which it follows as above that $\lim (r_n' - r_n) = 0$. This in turn gives that $\lim \phi'(r_n' - r_n) = 0$, so that $\lim \phi'(r_n) = \lim \phi'(r_n')$. This shows that the definition of $\omega(\hat{r})$ is not dependent upon the particular sequence $\{r_n\}$ chosen.

It is easily verified that ω is a ring-homomorphism. Suppose that $\hat{r} \in \hat{I}_k$. Since $\hat{I}_k$ is the closure of $\phi(I_k)$ in $\hat{R}$, there is a sequence $\{r_n\}$ of elements of I_k such that $\hat{r} = \lim \phi(r_n)$. Then

$$\omega(\hat{r}) = \lim \phi'(r_n) \in \hat{I}_k'.$$

Thus $\omega(\hat{I}_k) \subseteq \hat{I}_k'$ and

$$\omega(x + \hat{I}_k) \subseteq \omega(x) + \hat{I}_k'$$

for all $x \in \hat{R}$. This gives that ω is continuous.

We have now shown that there is a unique continuous mapping $\omega \colon \hat{R} \to \hat{R}'$ making the diagram Fig. 5.7 commutative. Further, ω is a ring-homomorphism and $\omega(\hat{I}_n) \subseteq \hat{I}_n'$. By a similar token, there is a unique continuous mapping $\omega' \colon \hat{R}' \to \hat{R}$ such that Fig. 5.8 is commutative, and it is clear from the constructions of ω and ω' that they are inverse mappings. Further, $\omega'(\hat{I}_n') \subseteq \hat{I}_n$. It follows that ω *is a ring-isomorphism and also a homeomorphism*;[†] *further*, $\omega(\hat{I}_n) = \hat{I}_n'$. This describes the extent to which the completion of a ring is unique. For suppose that $\phi \colon R \to \hat{R}$ is a completion of R and let $\omega \colon \hat{R} \to \hat{R}'$ be a ring-isomorphism. Put $\hat{I}_n' = \omega(\hat{I}_n)$. Then $\{\hat{I}_n'\}$ is a filtration on $\hat{R}'$ and ω is a homeomorphism. Further, $\omega\phi \colon R \to \hat{R}'$ is a completion of R.

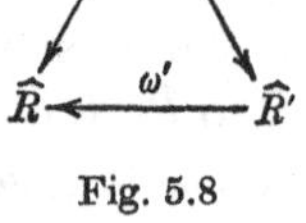

Fig. 5.8

---

† A mapping between topological spaces is a *homeomorphism* if it is a bijection, is continuous and has a continuous inverse.

Although we have established the essential uniqueness of ring completion, no assertions have as yet been made about the existence of completions of rings. Very general methods for the construction of completions are available (see, for example, *DGN Lessons* Chapter 9). We shall demonstrate the connection between injective modules and completions by using injective modules to construct the completion of a local ring.

Let R be a local ring with maximal ideal M. Then R is a topological ring under its natural topology, and this topology is Hausdorff, since $\bigcap_{n=1}^{\infty} M^n = 0$ (Proposition 4.23 Corollary 2). We put $E = E(R/M)$ and denote by H the ring of endomorphisms of E, i.e.

$$H = \mathrm{Hom}_R(E, E).$$

There is a mapping $\qquad \phi: R \to H$

given by $(\phi(r))(e) = re$, where $r \in R$, $e \in E$, and this mapping is a ring-homomorphism. In fact, by Proposition 2.26 Corollary 2,

$$\mathrm{Ker}\, \phi = \mathrm{Ann}_R E = 0,$$

so that ϕ *is an embedding*. We shall endow H with a filtration in such a way that (H, ϕ) is a completion of R. Already

$$\mathrm{Ker}\, \phi = \bigcap_{n=1}^{\infty} M^n.$$

Although Proposition 5.13 tells us that H is a commutative ring, we do not need to use this fact. Indeed, it will also follow from the fact that (H, ϕ) is a completion of R.

For each n, we put

$$A_n = 0:_E M^n, \quad H_n = 0:_H A_n.$$

Thus H_n consists of all endomorphisms of E which annihilate A_n. Clearly H_n is a left ideal of H. Suppose that $f \in H_n$, $g \in H$. Now $g(A_n) \subseteq A_n$, so that $fg(A_n) \subseteq f(A_n) = 0$. Thus, $fg \in H_n$ and H_n is a two-sided ideal of H. Further

$$H_1 \supseteq H_2 \supseteq H_3 \supseteq \dots,$$

so that $\{H_n\}$ is a filtration of H.

We now show that H is complete in the topology defined by this filtration. Since $\bigcup_{n=1}^{\infty} A_n = E$ (Proposition 4.23), it follows that $\bigcap_{n=1}^{\infty} H_n = 0$ and H is Hausdorff. Let $\{f_n\}$ be a Cauchy sequence of elements of H. For each k, there exists a positive integer $\nu(k)$ such that $f_m - f_{\nu(k)} \in H_k$ for every $m \geqslant \nu(k)$, i.e. f_m and $f_{\nu(k)}$ agree on A_k whenever $m \geqslant \nu(k)$. It may be assumed that, for all k, $\nu(k+1) \geqslant \nu(k)$. We now define $f: E \to E$ by

$$f(e) = f_{\nu(k)}(e) \quad \text{if} \quad e \in A_k.$$

Then $f \in H$ and $f_m - f \in H_k$ whenever $m \geqslant \nu(k)$. Thus the Cauchy sequence $\{f_n\}$ converges to f and H is complete.

We next show that $\phi(R)$ is dense in H. Let $f \in H$. By Proposition 5.12, there is for each positive integer k an element r_k of R such that $f(e) = r_k e$ for all $e \in A_k$, i.e. such that $f - \phi(r_k) \in H_k$. The sequence $\{\phi(r_k)\}$ of elements of $\phi(R)$ thus has limit f, and this shows that $\phi(R)$ is dense in H. Finally, by Proposition 4.23,

$$\phi(R) \cap H_n = \phi(0:A_n) = \phi(M^n).$$

This completes the verification that (H, ϕ) is a completion of R.

We have now established the following result:

THEOREM 5.14 *Let R be a local ring with maximal ideal M, let $E = E(R/M)$ and let H be the ring of endomorphisms of E. Give R its natural topology and H the topology arising from the filtration $\{H_n\}$ defined above. Then H is a completion of R.* $\square$

In the statement of Theorem 5.14, we have omitted to refer to the ring-homomorphism $\phi: R \to H$, although it is, of course, part of the completion.

COROLLARY 1 *Let R be a Noetherian ring and P a prime ideal of R. Then, with a suitable filtration, $\mathrm{Hom}_R(E(R/P), E(R/P))$ is a completion of the local ring R_P in its natural topology.*

Proof. See Propositions 5.4, 5.6 and 5.5 Corollary. $\square$

The next corollary to Theorem 5.14 indicates how the statement that a local ring is complete may be reformulated in purely algebraic terms; no reference to topological notions is made.

COROLLARY 2   *Let R be a local ring with maximal ideal M. Then the following statements are equivalent:*

(a)   *R is complete in its natural topology;*

(b)   *the mapping $\phi: R \to H$ is a ring-isomorphism;*

(c)   *every endomorphism of $E(R/M)$ consists of multiplication by an element of R.*

Proof. Suppose first that R is complete in its natural topology. Then, in the terminology of Theorem 5.14, R has completions (R, id) and (H, ϕ). By uniqueness, there is a ring-isomorphism $\omega: R \to H$ such that the diagram Fig. 5.9 is commutative. Clearly $\omega = \phi$ and (b) follows. Now assume (b). Then every endomorphism of $E(R/M)$ is of the form $\phi(r)$ for some $r \in R$. This is just the statement (c).

Fig. 5.9

Conversely, if (c) holds then the mapping ϕ is a surjection. Since ϕ is already an embedding, we have that (c) implies (b). Finally, assume (b). Then $\phi(M^n) = \phi(R) \cap H_n = H_n$ for all n. It follows that every Cauchy sequence of R has a limit and R is complete. $\square$

THEOREM 5.15   *Let R be a local ring with maximal ideal M, let $E = E(R/M)$ and let the completion H of R be given as in Theorem 5.14. Then H is a local ring with H_1 as its maximal ideal. Further, for each n, $H_n = H_1^n$, so that the topology defined on H by the filtration $\{H_n\}$ is its natural topology. Finally E, considered as an H-module, is isomorphic to an injective envelope of H/H_1.*

Proof. Since H is a completion of the commutative ring R, it is commutative. Also, by Proposition 3.12, H is quasi-local, and by Lemma 3.10 its non-units are precisely those elements f for which $\mathrm{Ker} f \neq 0$. Suppose that f is a non-unit of H. Then

$$(\mathrm{Ker} f) \cap (R/M) \neq 0,$$

so that $\mathrm{Ker}\, f \supseteq R/M = A_1$ (Lemma 4.24). Thus $f \in H_1$. Now $H_1 \neq H$, so that H_1 consists of all the non-units of H and H_1 is the maximal ideal of H.

Let I be any ideal of H. We shall show that I is finitely generated and hence that H is Noetherian. Put

$$A = 0:_E I = \bigcap_{f \in I} \mathrm{Ker} f.$$

By Theorem 4.30, E is Artinian, so the family consisting of all submodules of E of the form

$$\operatorname{Ker} f_1 \cap \operatorname{Ker} f_2 \cap \ldots \cap \operatorname{Ker} f_k, \qquad (5.3.2)$$

where $f_1, f_2, \ldots, f_k \in I$, has a minimal member. Suppose that (5.3.2) is a minimal member of this family. Then, for any $f \in I$,

$$\operatorname{Ker} f \cap \operatorname{Ker} f_1 \cap \ldots \cap \operatorname{Ker} f_k = \operatorname{Ker} f_1 \cap \ldots \cap \operatorname{Ker} f_k,$$

so that $\operatorname{Ker} f \supseteq \operatorname{Ker} f_1 \cap \operatorname{Ker} f_2 \cap \ldots \cap \operatorname{Ker} f_k$ and

$$A = \operatorname{Ker} f_1 \cap \operatorname{Ker} f_2 \cap \ldots \cap \operatorname{Ker} f_k. \qquad (5.3.3)$$

We define $\quad g: E \to E \oplus E \oplus \ldots \oplus E \quad$ (k terms)

by $g(e) = (f_1(e), f_2(e), \ldots, f_k(e))$, where $e \in E$. Then g is an R-homomorphism and has kernel A. Let $f \in I$. Then $f(A) = 0$ and f, g both give rise to induced mappings

$$f^*: E/A \to E, \quad g^*: E/A \to E \oplus E \oplus \ldots \oplus E.$$

Moreover g^* is a monomorphism. Consider the commutative diagram Fig. 5.10, where $\nu: E \to E/A$ is the natural mapping. Since E is injective, there is an R-homomorphism

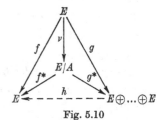

Fig. 5.10

$$h: E \oplus E \oplus \ldots \oplus E \to E$$

such that $hg^* = f^*$. For $1 \leqslant i \leqslant k$, denote by

$$\alpha_i: E \to E \oplus E \oplus \ldots \oplus E$$

the ith injection mapping. Then, for $e \in E$,

$$f(e) = f^*\nu(e) = hg^*\nu(e) = hg(e) = \sum_{i=1}^{k} h\alpha_i f_i(e).$$

Thus $f = \sum_{i=1}^{k} (h\alpha_i) f_i$ and

$$I = Hf_1 + Hf_2 + \ldots + Hf_k.$$

We next show that, for each n, $H_1^n = H_n$. Let $r_1, r_2, \ldots, r_s$ be a set of generators of M^n. Then

$$A_n = 0:_E M^n = \bigcap_{i=1}^{s} (0:_E r_i).$$

For $r \in R$, the element $\phi(r)$ of H is given by $(\phi(r))\,(e) = re$, where $e \in E$. Thus

$$A_n = \operatorname{Ker} \phi(r_1) \cap \operatorname{Ker} \phi(r_2) \cap \ldots \cap \operatorname{Ker} \phi(r_s).$$

Note that $\phi(r_i) \in H_n$ for $1 \leqslant i \leqslant s$, because $(\phi(r_i))\,(A_n) = r_i A_n = 0$. We are now in a situation parallel to that of (5.3.3), so that, if $\psi \in H_n$, we can deduce as before that

$$\psi \in H\phi(r_1) + H\phi(r_2) + \ldots + H\phi(r_s).$$

Thus $$H_n = H\phi(r_1) + H\phi(r_2) + \ldots + H\phi(r_s)$$

and $$H_n = \phi(M^n)\,H = \phi(M)^n\,H = (\phi(M)\,H)^n.$$

But the case $n = 1$ gives $H_1 = \phi(M)\,H$, so that $H_n = H_1^n$, as required.

The R-module E can also be regarded as a module over its ring of endomorphisms H, and every H-submodule of E is also an R-submodule, because $re = (\phi(r))\,(e)$ for $r \in R$, $e \in E$. On the other hand, let $e \in E$ and $f \in H$. Then, by Proposition 5.12 Corollary, there exists $r \in R$ such that $f(e) = re$. Thus every R-submodule is also an H-submodule. Hence the structure of E as an H-module is identical with its structure as an R-module. In particular, R/M is a simple H-submodule of E, so that $R/M \approx H/H_1$, this isomorphism being an H-isomorphism.

We shall show that E is an injective envelope of R/M when E and R/M are regarded as H-modules. Certainly E is an essential extension of R/M, so we need only show that E is injective as an H-module. Denote by E' an injective envelope of E as an H-module. Then E' is also an injective envelope of the H-module R/M; since $R/M \approx H/H_1$, this means that E' is an indecomposable H-module.

We can give E' the structure of an R-module by defining $re' = \phi(r)\,e'$ for $r \in R$ and $e' \in E'$. Then H-submodules of E' are also R-submodules. Now E is injective as an R-module, so there is an R-submodule F of E' such that $E' = E + F$ (d.s.). If we can show that F is also an H-submodule of E', then, since E' is an indecomposable H-module, we shall have $F = 0$ and $E' = E$, which will complete the proof.

Consider any non-zero element x of F and let $f \in H$. By Proposition 4.20 and Theorem 4.19 Corollary, $0:_H x$ is an H_1-primary ideal of H, so by Lemma 4.18 there is a positive integer l such that $H_l = H_1^l \subseteq 0:_H x$, whence $H_l x = 0$. Now $\phi(R)$ is dense in H, so there exists $r \in R$ such that $f - \phi(r) \in H_l$, whence

$$fx = \phi(r)x = rx \in F.$$

This shows that F is an H-submodule of E'. Everything is now proved. $\square$

5.4 Duality

Although the results of this section will be concerned principally with modules over complete local rings, we begin with some very general considerations.

Let R be a commutative ring and let E be an R-module. For an R-module N, we write

$$N^* = \mathrm{Hom}_R(N, E)$$

and call N^* the *dual* of N (relative to E). Then N^* has the structure of an R-module.

Consider a fixed R-module A. We use the symbol B to denote a typical submodule of A, and write

$$B^\lambda = \{f \in A^* : f(B) = 0\}.$$

Then B^λ is a submodule of A^*. Let B_1, B_2 be submodules of A. If $B_1 \subseteq B_2$, then $B_1^\lambda \supseteq B_2^\lambda$, so that the correspondence $B \to B^\lambda$ from the submodules of A to the submodules of A^* reverses inclusions. Further,

$$(B_1 + B_2)^\lambda = B_1^\lambda \cap B_2^\lambda$$

and
$$(IB)^\lambda = B^\lambda :_{A^*} I, \qquad (5.4.1)$$

where I denotes an ideal of R.

Now let K denote a typical submodule of A^*. We write

$$K^\mu = \{a \in A : f(a) = 0 \text{ for all } f \in K\}.$$

Then K^μ is an R-submodule of A. Let K_1, K_2 be submodules of A^*. If $K_1 \subseteq K_2$, then $K_1^\mu \supseteq K_2^\mu$, so that the correspondence

$K \to K^\mu$, from the submodules of A^* to the submodules of A, reverses inclusions. Further,

$$(K_1 + K_2)^\mu = K_1^\mu \cap K_2^\mu$$

and
$$(IK)^\mu = K^\mu :_A I.$$

We now have correspondences $B \to B^\lambda$ and $K \to K^\mu$, the former from the submodules of A to the submodules of A^*, the latter in the reverse direction. It is natural to wonder when these are inverses of one another. Clearly $B \subseteq B^{\lambda\mu}$, so that $B^\lambda \supseteq B^{\lambda\mu\lambda}$. Also, $K \subseteq K^{\mu\lambda}$, so that $B^\lambda \subseteq B^{\lambda\mu\lambda}$. Hence $B^\lambda = B^{\lambda\mu\lambda}$, and similarly $K^\mu = K^{\mu\lambda\mu}$.

Suppose now that E is an injective cogenerator of R, so that E is an injective R-module and, given a non-zero R-module X, there is a non-zero R-homomorphism $X \to E$. Consider any B such that $B \subset B^{\lambda\mu}$. Then there is a non-zero R-homomorphism $\theta: B^{\lambda\mu}/B \to E$, which can be extended to an R-homomorphism $\phi: A/B \to E$ since E is injective. Let $\nu: A \to A/B$ be the natural mapping. Then $\phi\nu \in A^*$ and $\phi\nu(B) = 0$, so that $\phi\nu \in B^\lambda$. But $\phi\nu(B^{\lambda\mu}) \neq 0$, so that $\phi\nu \notin B^{\lambda\mu\lambda}$. This contradicts the fact that $B^\lambda = B^{\lambda\mu\lambda}$. Thus, *when E is an injective cogenerator of R*,

$$B = B^{\lambda\mu} \text{ for all } B.$$

This tells us that, in this situation, the correspondence $B \to B^\lambda$ is an injection and $K \to K^\mu$ is a surjection. However, we cannot in general say that $K = K^{\mu\lambda}$ for all K, although we shall meet a situation when this is indeed the case. *When $K = K^{\mu\lambda}$ for all K*, our two correspondences are bijections which are inverses of each other, and

$$(B_1 + B_2)^\lambda = B_1^\lambda \cap B_2^\lambda, \quad (IB)^\lambda = B^\lambda :_{A^*} I,$$
$$(B_1 \cap B_2)^\lambda = B_1^\lambda + B_2^\lambda, \quad (B :_A I)^\lambda = IB^\lambda,$$

where B, B_1, B_2 are submodules of A and I is an ideal of R.

Suppose we take the module A to be the ring R, whilst still assuming that E is an injective cogenerator of R (but not that $K = K^{\mu\lambda}$ for all K). The dual R^* is isomorphic to E; under this isomorphism, $f \in R^*$ corresponds to $f(1) \in E$. We thus have an injection from the collection of ideals of R to the collection of submodules of E. This injection is given by

$$I \to 0 :_E I, \tag{5.4.2}$$

where I denotes a typical ideal of R. We also have a surjection from the submodules of E to the ideals of R given by

$$C \to 0:C, \qquad (5.4.3)$$

where C denotes a typical submodule of E. If we begin with the ideal I, form the submodule $0:_E I = C$ and then form $0:C$, we return to the original ideal I.

We recall that, when R is a quasi-local ring with maximal ideal M, then $E(R/M)$ is an injective cogenerator of R.

PROPOSITION 5.16   *Let R be a commutative quasi-local ring with maximal ideal M, let A be an R-module, let $E = E(R/M)$ and let the notation be as above.*

(i)   *When B is a non-zero cyclic submodule of A, then B^λ is an irreducible M-isotopic submodule of $A*$.*

(ii)   *Suppose further that R is a complete local ring. Then, when B is an irreducible M-isotopic submodule of A, B^λ is a non-zero cyclic submodule of $A*$.*

Proof. (i) Put $B = Ra$, where $a \neq 0$, and define $\phi: A* \to E$ by $\phi(f) = f(a)$, where $f \in A*$. Then ϕ is an R-homomorphism and its kernel is B^λ. Hence ϕ induces a monomorphism $A*/B^\lambda \to E$. Moreover $B \neq 0$, whence $B^\lambda \neq 0^\lambda = A*$. It follows from Proposition 2.28 that

$$E(A*/B^\lambda) \approx E = E(R/M),$$

whence B^λ is an irreducible M-isotopic submodule of $A*$.

(ii) Now suppose that R is a complete local ring and let B be an irreducible M-isotopic submodule of A. Then there is an isomorphism $E(A/B) \approx E(R/M) = E$ and so a monomorphism $\phi: A/B \to E$. Consider an element f of B^λ, so that $f: A \to E$ and $f(B) = 0$. Then f induces an R-homomorphism $g: A/B \to E$, and we have the commutative diagram Fig. 5.11, where ν is the natural mapping. Since ϕ is a monomorphism and E is injective, there is an R-homomorphism $\psi: E \to E$ such that $\psi\phi = g$. Then $f = g\nu = \psi\phi\nu$. Since R is complete, there is an element r of R such that $\psi = r \, \mathrm{id}_E$ (Theorem

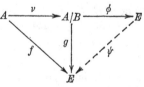

Fig. 5.11

5.14 Corollary 2). Thus $f = r(\phi\nu)$, which shows that B^λ is generated by the single element $\phi\nu$. Note that, since $B \neq A$, we have $B^\lambda \neq A^\lambda = 0.\square$

For the rest of this section, R will denote a complete local ring with maximal ideal M and E will denote $E(R/M)$. An R-homomorphism $A \to B$ gives rise to an R-homomorphism $B^* \to A^*$, and hence to an R-homomorphism $A^{**} \to B^{**}$. Since E is injective, an exact sequence

$$0 \to A \to B \to C \to 0$$

gives rise to the exact sequence

$$0 \to C^* \to B^* \to A^* \to 0$$

and thence to the exact sequence

$$0 \to A^{**} \to B^{**} \to C^{**} \to 0.$$

For each R-module A, there is an R-homomorphism

$$\phi_A \colon A \to A^{**}$$

given by $(\phi_A(a))(a^*) = a^*(a)$, where $a \in A$ and $a^* \in A^*$. Moreover, if $A \to B$ is a homomorphism of R-modules, then Fig. 5.12 is commutative. We describe this situation by saying that ϕ_A is a *natural homomorphism*. Further, *each ϕ_A is a monomorphism.* For consider a non-zero element a of A. Since E is an injective cogenerator of R, there is an $a^* \in A^*$ such that $a^*(a) \neq 0$. Thus $\phi_A(a) \neq 0$.

Fig. 5.12

Fig. 5.13

We shall now identify the mapping ϕ_R. We have the diagram Fig. 5.13, where ϕ is given by $(\phi(r))(e) = re$ $(r \in R, e \in E)$. To describe τ, we recall that $R^* \approx E$, where $r^* \in R^*$ corresponds to $r^*(1)$. This gives rise to the isomorphism $\tau \colon E^* \to R^{**}$, so that

$$(\tau(e^*))(r^*) = e^*(r^*(1)) \quad (e^* \in E^*, r^* \in R^*).$$

Now

$$((\tau\phi)(r))(r^*) = (\tau(\phi(r)))(r^*) = (\phi(r))(r^*(1)) = rr^*(1)$$
$$= r^*(r)$$
$$= (\phi_R(r))(r^*),$$

so that $\tau\phi = \phi_R$ and Fig. 5.13 is commutative. Now ϕ is just the canonical mapping of R into its completion. Since R is assumed to be complete, ϕ is an isomorphism (Theorem 5.14 Corollary 2). Hence ϕ_R *is an isomorphism.*

We have a commutative diagram Fig. 5.14. Further, the mapping $\phi_R^*: R^{***} \to R^*$ induced by ϕ_R is an isomorphism. We assert that

$$\phi_{R^*} = (\phi_R^*)^{-1}.$$

Consider $r^* \in R^*$, $r \in R$. Then

$$((\phi_R^* \phi_{R^*})(r^*))(r) = (\phi_{R^*}(r^*))(\phi_R(r))$$
$$= (\phi_R(r))(r^*)$$
$$= r^*(r).$$

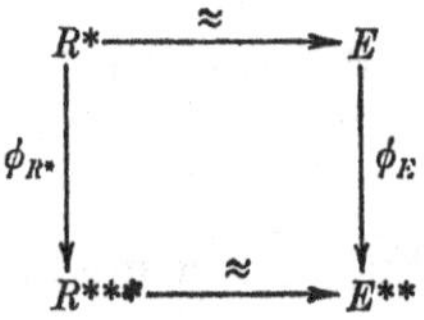

Fig. 5.14

Hence $\phi_R^* \phi_{R^*} = \mathrm{id}$, so that $\phi_{R^*} = (\phi_R^*)^{-1}$, since ϕ_R^* is an isomorphism. Thus ϕ_{R^*} is an isomorphism, and so ϕ_E *is an isomorphism.*

Now let A, B be R-modules, and consider the exact sequence

$$0 \to A \to A \oplus B \to B \to 0,$$

where the mappings are the injection and projection mappings. This gives rise to the exact sequence

$$0 \to A^{**} \to (A \oplus B)^{**} \to B^{**} \to 0,$$

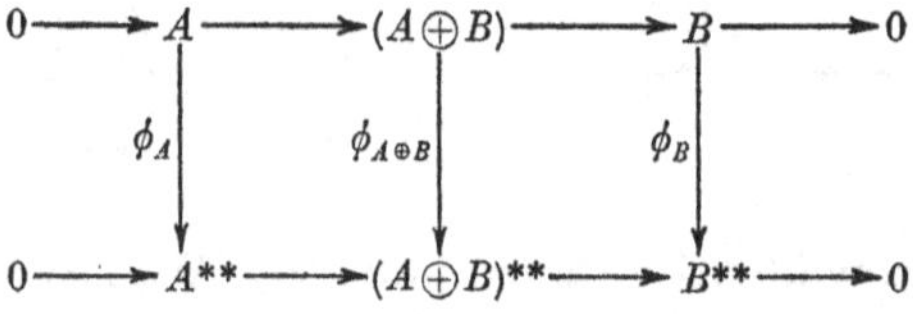

Fig. 5.15

and Fig. 5.15 is commutative. It follows from Lemma 5.9 that, if ϕ_A and ϕ_B are isomorphisms, then $\phi_{A \oplus B}$ is an epimorphism, and so also an isomorphism. For a non-negative integer k, we denote by A^k the direct sum of k copies of A. It follows by induction from the above that, *for every non-negative integer k, ϕ_{R^k} and ϕ_{E^k} are isomorphisms.*

LEMMA 5.17 *If A is either a Noetherian or an Artinian R-module, then ϕ_A is an isomorphism.*

Proof. Suppose first that A is a Noetherian R-module. Then A is a homomorphic image of R^k for some non-negative integer k, and we have an exact sequence of the form

$$R^k \to A \to 0.$$

This gives the commutative diagram Fig. 5.16 with exact rows. Since ϕ_{R^k} is an isomorphism, it follows that ϕ_A is an epimorphism and so also an isomorphism.

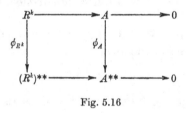

Fig. 5.16

Now suppose that A is an Artinian R-module. Then A is finitely embedded (Theorem 3.21), so there is an exact sequence of the form
$$0 \to A \to E^k \to B \to 0$$

for some non-negative integer k. This gives the commutative diagram Fig. 5.17 where the rows are exact. Since ϕ_{E^k} is an isomorphism, Lemma 5.10 gives that ϕ_A is an epimorphism and hence also an isomorphism.□

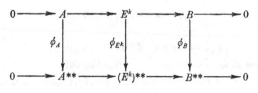

Fig. 5.17

LEMMA 5.18 *Let A be a Noetherian R-module. Then A^* is Artinian.*

Proof. Since A is Noetherian, it is a homomorphic image of R^k for some non-negative integer k, and we have an exact sequence of the form

$$R^k \to A \to 0.$$

This gives the exact sequence

$$0 \to A^* \to (R^k)^*.$$

But $(R^k)^* \approx (R^*)^k$ by Proposition 1.24, and $R^* \approx E$. Hence $(R^k)^* \approx E^k$. Now E is finitely embedded and so Artinian (Theorem 4.30). Hence E^k, and so also A^*, is Artinian. □

Note that the local ring R did not need to be complete for Lemma 5.18.

LEMMA 5.19 *Let A be an Artinian R-module. Then A^* is Noetherian.*

Proof. By Theorem 3.21, A is finitely embedded, so that there is an exact sequence of the form

$$0 \to A \to E^k$$

for some non-negative integer k. This gives rise to the exact sequence

$$(E^k)^* \to A^* \to 0.$$

But $(E^k)^* \approx (E^*)^k$. Further, R is complete, so $E^* \approx R$ (Theorem 5.14 Corollary 2). Thus $(E^k)^* \approx R^k$, which is Noetherian. Hence A^* is Noetherian. □

We collect together the results of Lemmas 5.17, 5.18 and 5.19 in the next theorem.

THEOREM 5.20 *Let R be a complete local ring with maximal ideal M, let $E = E(R/M)$, let A be an R-module and let*

$$A^* = \mathrm{Hom}_R(A, E).$$

Then:

(i) *when A is Noetherian, A^* is Artinian;*

(ii) *when A is Artinian, A^* is Noetherian;*

(iii) *when A is either Noetherian or Artinian, A and A^{**} are naturally isomorphic.* □

We may describe this result using the language of categories. Let R be a complete local ring with maximal ideal M and let $E = E(R/M)$. Denote by $*$ the functor $\text{Hom}_R(.,E)$ from the category of R-modules into itself. Then $*$ defines contravariant functors (1) from the category of Noetherian R-modules to the category of Artinian R-modules, (2) from the category of Artinian R-modules to the category of Noetherian R-modules. Further, (3) these two categories are equivalent under $*$. Here, (1), (2) and (3) are just restatements of (i), (ii) and (iii) respectively in Theorem 5.20. This describes the duality that exists between Noetherian and Artinian modules over a complete local ring.

In the next result we describe a situation in which the correspondence $B \to B^\lambda$ from the submodules of an R-module A to the submodules of its dual is a bijection, with inverse $K \to K^\mu$. It is already known that $B^{\lambda\mu} = B$ for all B, so that $B \to B^\lambda$ is an injection and $K \to K^\mu$ a surjection.

THEOREM 5.21 *Let R be a complete local ring with maximal ideal M, let $E = E(R/M)$, let A be an R-module which is either Noetherian or Artinian and let $A^* = \text{Hom}_R(A,E)$. Then there is a one-one correspondence between the submodules B of A and the submodules of A^* given by $B \leftrightarrow B^\lambda$, and*

$$(B_1 + B_2)^\lambda = B_1^\lambda \cap B_2^\lambda, \quad (IB)^\lambda = B^\lambda :_{A^*} I,$$
$$(B_1 \cap B_2)^\lambda = B_1^\lambda + B_2^\lambda, \quad (B :_A I) = IB^\lambda,$$

where B_1, B_2 are also submodules of A and I is an ideal of R. Moreover, under this correspondence, the non-zero cyclic submodules of A correspond to the irreducible M-isotopic submodules of A^*, and the irreducible M-isotopic submodules of A correspond to the non-zero cyclic submodules of A^*.

Proof. The first part will follow if we can show that the correspondence $K \to K^\mu$ is an injection. There is an injection $K \to K^{\lambda'}$ from the submodules of A^* to the submodules of A^{**}, where
$$K^{\lambda'} = \{a^{**} \in A^{**} : a^{**}(K) = 0\}.$$

But, by Lemma 5.17, we have the natural isomorphism
$$\phi_A : A \to A^{**}$$

given by

$$(\phi_A(a))\,(a^*) = a^*(a), \text{ where } a \in A \text{ and } a^* \in A^*,$$

and $$\phi_A(K^\mu) = K^{\lambda'}.$$

But now the correspondence $K \to \phi_A^{-1}(K^{\lambda'}) = K^\mu$ is an injection, as required.

We have seen in Proposition 5.16 that $B \to B^\lambda$ gives an injection from the non-zero cyclic (resp. irreducible M-isotopic) submodules of A to the irreducible M-isotopic (resp. non-zero cyclic) submodules of A^*. Let K be an irreducible M-isotopic (resp. non-zero cyclic) submodule of A^*. Then, by Proposition 5.16, $K^{\lambda'}$ is a non-zero cyclic (resp. irreducible M-isotopic) submodule of A^{**}. The same statements now hold for K^μ as a submodule of A. But $K = K^{\mu\lambda}$. The remaining assertions of Theorem 5.21 now follow. $\square$

The corollary which follows is essentially the special case of Theorem 5.21 when $A = R$. We recall that $R^* \approx E$ and that we now have correspondences $I \to 0 :_E I$ and $C \to 0 : C$ from the ideals I of R to the submodules C of E and vice versa (see (5.4.2), (5.4.3)).

COROLLARY *Let R be a complete local ring with maximal ideal M and let $E = E(R/M)$. Then there is a one–one correspondence between the ideals I of R and the submodules C of E, given alternatively by $I \leftrightarrow 0 :_E I$ or $0 : C \leftrightarrow C$. Moreover, under this correspondence, the non-zero principal ideals of R correspond to the irreducible M-isotopic submodules of E, and the irreducible M-primary ($= M$-isotopic)† ideals of R correspond to the non-zero cyclic submodules of E.* $\square$

In this corollary, the result that there is a bijection from the non-zero cyclic submodules of E to the irreducible M-primary ideals of R, given by $C \to 0 : C$, does not need the local ring R to be complete. We shall demonstrate this by means of an ad hoc argument. If $C = Re$ is a non-zero cylic submodule of E, then, by Theorem 4.19 Corollary and Proposition 4.20, $0 : C = 0 : e$ is an irreducible M-primary ideal of R; and every irreducible M-primary ideal of R arises in this way. It remains for us to show that, when x, y are non-zero elements of E such that $0 : x = 0 : y$,

---

† See Theorem 4.19 Corollary.

then $Rx = Ry$. This will follow as a corollary to the following lemma:

LEMMA 5.22 *Let R be a local ring with maximal ideal M, let $E = E(R/M)$ and let x, y be non-zero elements of E such that $0:x \subseteq 0:y$. Then $Ry \subseteq Rx$.*

Proof. Since $0:x \subseteq 0:y$, there is a well-defined R-homomorphism

$$\theta: Rx \to Ry$$

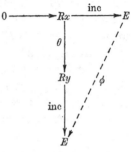

given by $\theta(sx) = sy$, where $s \in R$. Consider Fig. 5.18. There is an endomorphism ϕ of E which makes the resulting diagram commutative. Proposition 5.12 Corollary now shows that there is an element r of R such that $\phi(x) = rx$. Thus $y = rx$ and $Ry \subseteq Rx$.☐

Fig. 5.18

Exercises on Chapter 5

5.1 Show that an Artinian local ring is complete.

5.2 Let R be a complete local domain with maximal ideal M and let m be a non-zero element of M. Let $\{x_n\}$, $\{y_n\}$ be sequences whose elements are either 0 or 1 in R. Show that $\sum_{n=1}^{\infty} x_n m^n$ and $\sum_{n=1}^{\infty} y_n m^n$ are defined in R and that

$$\sum_{n=1}^{\infty} x_n m^n = \sum_{n=1}^{\infty} y_n m^n$$

if and only if $x_n = y_n$ for every n. Deduce that a complete local domain is uncountable.

Show that the ring Z_p consisting of all rational numbers of the form a/b, where $p \nmid b$, is a local domain which is not complete.

5.3 Let R be a commutative ring, let M be a maximal ideal of R and let $\phi_M: R \to R_M$ be the canonical ring-homomorphism. Prove that R/M is an injective R-module if and only if ϕ_M induces a ring-isomorphism $R/M \approx R_M$. Deduce that R is a regular ring if and only if ϕ_M induces such an isomorphism for every maximal ideal M of R. [*Hint:* Use Proposition 5.5 and Exercise 2.14]

5.4 Let R be a complete local ring with maximal ideal M and let A, B be submodules of $E(R/M)$. Show that every R-homomorphism from A to B consists of multiplication by an element of R.

5.5 Let R be a commutative ring and let E be an injective cogenerator of R. Suppose that E is Artinian (resp. Noetherian). Show that R is a Noetherian (resp. Artinian) ring.

5.6 Let R be a commutative quasi-local ring with maximal ideal M, put $E = E(R/M)$, let A be an R-module and put $A^* = \mathrm{Hom}_R(A, E)$.

(i) Let B be a finitely generated submodule of A, and suppose that $\{a_1, a_2, ..., a_k\}$ is a minimal set of generators of B. Show that $B^\lambda = \{f \in A^* : f(B) = 0\}$ is an irredundant intersection of k irreducible M-isotopic submodules of A^*. [*Hint:* Use Proposition 2.24]

(ii) Let R be a complete local ring and let A be either Noetherian or Artinian. Let C be an irredundant intersection of l irreducible M-isotopic submodules of A. Show that C^λ is a finitely generated submodule of A^* and the minimal number of elements required to generate C^λ is l.

5.7 Let $P_1, P_2, ..., P_k$ $(k \geqslant 1)$ be prime ideals of a commutative ring R and let I be an ideal of R such that $I \subseteq P_1 \cup P_2 \cup ... \cup P_k$. Show that $I \subseteq P_i$ for some i.

5.8 Let R be a commutative Noetherian ring, let I be an ideal of R and let A be an R-module such that $E(A)$ is the direct sum of a finite number of its indecomposable injective submodules. Show that $0 :_A I \neq 0$ if and only if $0 :_A r \neq 0$ for every $r \in I$. [*Hint:* Use Exercise 5.7]

5.9 Let R be a complete local ring and let A be an R-module which is either Noetherian or Artinian. Let I be an ideal of R such that $IA = A$. Show that there is an element $r \in I$ such that $rA = A$. [*Hint:* Use Exercise 5.8]

5.10 Let R be a commutative quasi-local ring with maximal ideal M, let $E = E(R/M)$, let A be an R-module and let

$$A^* = \mathrm{Hom}_R(A, E).$$

Suppose that the ring R has the property that every endomorphism of E consists of multiplication by an element of R. Show that:

(i) when A is a submodule of a finitely generated R-module, then A^* is a homomorphic image of a finitely embedded R-module;

(ii) when A is a homomorphic image of a finitely embedded R-module, then A^* is a submodule of a finitely generated R-module;

(iii) when A is either a submodule of a finitely generated R-module or a homomorphic image of a finitely embedded R-module, then A and A^{**} are naturally isomorphic. (This generalizes Theorem 5.20.)

5.11 Let the ring R be as in Exercise 5.10, let A be an R-module which is either a submodule of a finitely generated R-module or a homomorphic image of a finitely embedded R-module, and let $A^* = \operatorname{Hom}_R(A, E)$. Establish a one–one correspondence between the submodules of A and the submodules of A^* under which the non-zero cyclic submodules of A correspond to the irreducible M-isotopic submodules of A^*, and the irreducible M-isotopic submodules of A correspond to the non-zero cyclic submodules of A^*. (This generalizes Theorem 5.21. Theorem 5.21 Corollary also generalizes to apply to these rings.)

Notes on Chapter 5

The methods used in the localization of a ring are due to E. Matlis. Localization does not stop at the ring itself and may also be extended to modules (see for example [22]).

The result that the ring of endomorphisms of $E(R/P)$ is the completion of R_P, where R is a Noetherian ring and P is a prime ideal of R, is due to E. Matlis (Theorem 5.14 and in particular Corollary 1). Our proof of Theorem 5.14 is based on the methods of P. Gabriel [9].

All the proofs previously known to us of the result that the completion of a local ring is Noetherian use graded rings. The proof of this result given in Theorem 5.15 is believed to be new.

There are some quite deep results about rings of endomorphisms of injective modules over general rings. For example, let E be an injective module, let H be the ring of endomorphisms of

E and let J be the Jacobson radical of H. Then H/J is a regular ring and idempotents lift modulo J (see C. Faith [5] Chapter 5). The structure of J is known (see Exercise 3.12).

The results on duality are due to E. Matlis. As in Exercises 5.10 and 5.11, a duality theory can be developed for commutative quasi-local rings R for which every endomorphism of $E(R/M)$ consists of multiplication by an element of R (M is the maximal ideal of R). Examples of such rings are complete local rings and also maximal valuation domains (see E. Matlis [20]). There are other classes of rings with this property.

Possibly the grandfather of results on duality is the well-known duality for finite-dimensional vector spaces. It is important to notice that finite dimensionality is necessary since the second dual of an infinite dimensional vector space is considerably larger than the original. One way to remedy this is to introduce a topology. This approach leads to Hilbert spaces (every Hilbert space is naturally isomorphic to its second dual, i.e. it is reflexive) and to the Pontryagin duality between Abelian groups and compact Abelian groups (see for example [8]). Injective cogenerators also induce dualities in categories other than categories of modules, e.g. Stone's duality between Boolean algebras and totally disconnected compact Hausdorff spaces (see [12]).

A quasi-Frobenius ring is one of the many possible generalizations of a semi-simple ring. It is a left and right Noetherian ring which is left self-injective, i.e. it is injective as a left module over itself. Such a ring is also right self-injective and left and right Artinian. (See Exercise 6.1 for commutative quasi-Frobenius rings.) There is a duality between the left Noetherian and the right Noetherian modules over a quasi-Frobenius ring. The fact that the group algebra of a finite group over a field is a quasi-Frobenius ring which is not in general a semi-simple ring is one reason why these rings are important. We refer the reader to [13].

6. Direct sum decompositions

6.1 Direct sums of cyclic modules

A module M over a division ring D is the direct sum of a family of submodules each isomorphic to D, and the cardinality of this family is an invariant of M, called its *dimension*. Moreover, two modules over the same division ring are isomorphic if and only if they have the same dimension. More generally, we may consider a module M over a semi-simple ring R. By Proposition 3.7, M is itself semi-simple and so is a direct sum of a family of simple submodules, say

$$M = \sum_{i \in I} S_i \quad \text{(d.s.)}.$$

We may ascribe invariants to M as follows. Let $\{T_j\}_{j \in J}$ be a representative family of non-isomorphic simple R-modules (so that each simple R-module is isomorphic to some T_j and $T_j \not\cong T_{j'}$ whenever $j \neq j'$). For each $j \in J$, we attach to M the cardinality of the family of those S_i which are isomorphic to T_j. We thus obtain a family of cardinal numbers indexed by J; it is a consequence of Theorem 3.13 Corollary that this family is an invariant of M. Again, two modules over the same semi-simple ring are isomorphic if and only if the corresponding families of cardinal numbers are the same. Of course, if we insist that every module over a ring R shall be a direct sum of simple modules, then R must be semi-simple.

One way of relaxing the requirement that every R-module shall be a direct sum of simple modules is to replace the word 'simple' by 'cyclic'. The primary object of this chapter is to discuss the following question: given a ring R, can every R-module be decomposed in a unique way as a direct sum of cyclic modules? We shall present a number of results in this direction; these will involve the restriction of either the class of rings or the class of modules considered. Since results of this type for modules over non-commutative rings are few and hard to obtain, we shall confine ourselves to commutative rings. Thus, *for the rest of this chapter, all rings are assumed to be commutative.*

Our first result provides another characterization of Artinian rings.

PROPOSITION 6.1 *Let R be a Noetherian ring. Then the following statements are equivalent:*

(a) *R is Artinian;*

(b) *every indecomposable injective R-module is Noetherian.*

Proof. Assume that R is Artinian and let E be an indecomposable injective R-module. By Theorem 4.5, $E \approx E(R/M)$ for some maximal ideal M of R. Hence E is finitely embedded and so Artinian (Theorem 4.30). Theorem 3.25 now gives that E is Noetherian.

Conversely, assume (b) and let P be an arbitrary prime ideal of R. By Theorem 4.6, to show that R is Artinian it suffices to show that P is maximal. Consider any element t of R, $t \notin P$. By Lemma 5.2, the mapping $\phi(t): E(R/P) \to E(R/P)$ given by $(\phi(t))(e) = te$, where $e \in E(R/P)$, is an isomorphism. Consider the collection of submodules $\phi(t)^{-k}(R/P)$ $(k = 0, 1, 2, \ldots)$ of $E(R/P)$. Since $E(R/P)$ is indecomposable (Lemma 2.29), our assumption gives that $E(R/P)$ is Noetherian. Hence this family of submodules has a maximal member $\phi(t)^{-l}(R/P)$ (say). An arbitrary element of $\phi(t)^{-l}(R/P)$ is of the form

$$\phi(t)^{-l}(e) = \phi(t)^{-l-1}(te),$$

where $e \in R/P$, so that $\phi(t)^{-l}(R/P) \subseteq \phi(t)^{-l-1}(R/P)$. It follows that $\phi(t)^{-l}(R/P) = \phi(t)^{-l-1}(R/P)$, whence

$$t(R/P) = (\phi(t))(R/P) = R/P.$$

Thus there exist $r \in R$ and $p \in P$ such that $1 = rt + p$, and $R = Rt + P$. Since this is true for every element t of R such that $t \notin P$, it follows that P is maximal. $\square$

LEMMA 6.2 *Let R be a quasi-local ring with maximal ideal M and let U be an R-module with the property that every finitely generated submodule of U is cyclic. Then the submodules of U are totally ordered.*

Proof. Suppose that there are submodules A, B of U such that $A \nsubseteq B$ and $B \nsubseteq A$. Then there exist elements a, b such that $a \in A$, $a \notin B$ and $b \in B$, $b \notin A$. The submodule $Ra + Rb$ of U is cyclic, say
$$Ra + Rb = Rc.$$

Note that $c \neq 0$. There exist elements $p, q, r, s \in R$ such that
$$a = pc, \quad b = qc, \quad c = ra + sb.$$

This gives $(1 - rp - sq)\, c = 0$. Now $1 - rp - sq$ cannot be a unit of R because $c \neq 0$, so that $1 - rp - sq \in M$. Hence either $rp \notin M$ or $sq \notin M$. If $rp \notin M$, then p is a unit and $b \in Rc = Ra \subseteq A$. Similarly, if $sq \notin M$, then $a \in B$. Either way we have a contradiction, so that one of $A \subseteq B$ and $B \subseteq A$ must hold, i.e. the submodules of U are totally ordered.□

Corollary   *Let R be a quasi-local ring with maximal ideal M and let $E = E(R/M)$. Suppose that every finitely generated submodule of E is cyclic. Then the ideals of R are totally ordered.*

Proof. We use the correspondences between the ideals of R and the submodules of E described in (5.4.2) and (5.4.3). Let I, J be ideals of R and put $A = 0:_E I$, $B = 0:_E J$. Then A, B are submodules of E and $I = 0:A$, $J = 0:B$. Now either $A \subseteq B$ or $B \subseteq A$, so either $I \supseteq J$ or $J \supseteq I$.□

Lemma 6.3   *Let R be a local ring. Then the following statements are equivalent:*

(a) *R is a principal ideal ring;*

(b) *there is an element p of R such that every non-zero ideal of R is of the form Rp^k, where k is a non-negative integer;*†

(c) *the ideals of R are totally ordered.*

Proof. Assume (a) and let Rp be the maximal ideal of R. Consider a non-zero ideal Rr of R. By Proposition 4.23 Corollary 2, $\bigcap_{n=1}^{\infty} Rp^n = 0$, so there is a non-negative integer k such that $Rp^k \supseteq Rr$ but $Rp^{k+1} \nsupseteq Rr$. We can thus write $r = sp^k$, where $s \in R$ but $s \notin Rp$. Thus s is a unit of R and $Rr = Rp^k$.

Since it is obvious that (b) implies (c), we now assume (c) and deduce (a). It is sufficient to consider an arbitrary non-zero ideal

† We define $p^0 = 1$.

I of R. Consider the family of all ideals properly contained in I. This family is non-empty so, since R is Noetherian, the family has a maximal member J (say). Let r be any element of I not in J. Since the ideals of R are totally ordered, $J \subset Rr$. Also $Rr \subseteq I$. The only possibility is that $I = Rr$. □

DEFINITION   *A ring R is said to be 'self-injective' if it is injective as a module over itself.*
We note that every ring which is isomorphic to a self-injective ring is itself self-injective. Also, a trivial ring is self-injective.

LEMMA 6.4   *Let R be a local Artinian principal ideal ring. Then, for every ideal I of R, the residue class ring R/I is self-injective.*

Proof. We first prove that the ring R is self-injective. Consider Fig. 6.1 where A is an ideal of R and $f: A \to R$ is an R-homomorphism. By Lemma 6.3, there is an element p of R for which the non-zero ideals of R are precisely $R, Rp, Rp^2, \ldots$, and

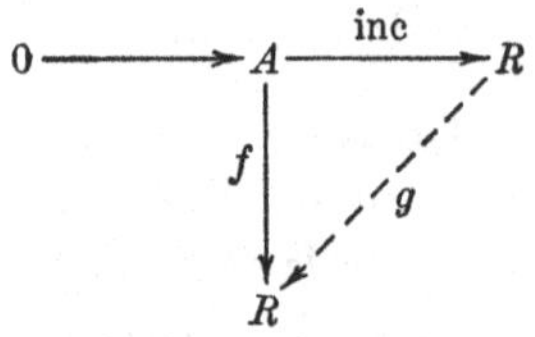

Fig. 6.1

$$\bigcap_{n=1}^{\infty} Rp^n = 0.$$

But R is Artinian. It follows that there is an integer s for which $p^s = 0$ but $p^{s-1} \neq 0$. Then the ideals of R are precisely $R, Rp, Rp^2, \ldots, Rp^s = 0$, and these are all distinct. We put $A = Rp^k$, where $k \leqslant s$. Suppose that Rp^l is the smallest ideal of R to which $f(p^k)$ belongs. Then we can write $f(p^k) = up^l$ for some unit u of R. Now $p^{s-k}f(p^k) = 0$, so that $p^{s-k+l} = 0$ and $l \geqslant k$. We now define the R-homomorphism $g: R \to R$ by $g(r) = rup^{l-k}$, where $r \in R$. This mapping completes our diagram and shows that R is self-injective.

Now consider an ideal I of R. If $I = R$, then certainly the ring R/I is self-injective. If $I \neq R$, then R/I is also a local Artinian principal ideal ring so, by what we have just proved, R/I is self-injective. □

LEMMA 6.5   *Let R be a quasi-local ring such that the ring R/I is self-injective for every ideal I of R. Then the ideals of R are totally ordered.*

Proof. Let M be the maximal ideal of R and consider a proper ideal I of R. The ring R/I is quasi-local with maximal ideal M/I. Now R/I regarded as a module over itself is indecomposable. For suppose that

$$R/I = (I_1/I) + (I_2/I) \quad \text{(d.s.)},$$

where I_1 and I_2 are ideals of R containing I. Then $R = I_1 + I_2$ and either $I_1 \nsubseteq M$ or $I_2 \nsubseteq M$, i.e. either $I_1 = R$ or $I_2 = R$. Thus R/I is indecomposable and injective as a module over itself, so by Proposition 2.28 its zero submodule is irreducible. This means that I is an irreducible ideal of R.

Now let A, B be proper ideals of R. Then $A \cap B$ is an irreducible ideal, so either $A \cap B = A$ or $A \cap B = B$, i.e. either $A \subseteq B$ or $B \subseteq A$. Thus the proper ideals of R are totally ordered.□

If we put together the previous three lemmas, we obtain the next lemma.

LEMMA 6.6 *Let R be a local Artinian ring. Then the following statements are equivalent:*

(a) *R is a principal ideal ring;*

(b) *there is an element p of R and a postive integer s such that the ideals of R are precisely $R, Rp, Rp^2, \ldots, Rp^s = 0$, these being all distinct;*

(c) *the ideals of R are totally ordered;*

(d) *for every ideal I of R, the ring R/I is self-injective.*□

In the next result, we shall describe all commutative rings R with the property that every R-module is a direct sum of cyclic submodules.

THEOREM 6.7 *For a commutative ring R, the following statements are equivalent:*

(a) *every R-module is a direct sum of cyclic submodules;*

(b) *R is an Artinian principal ideal ring;*

(c) *R is Noetherian and, for every ideal I of R, the ring R/I is self-injective.*

Proof. We first notice that each of the conditions (a), (b), (c) implies that R is an Artinian ring. For assume (a). Then the remark following Theorem 4.4 gives that R is Noetherian. Also, (a) gives that every indecomposable injective R-module is cyclic, and so Noetherian. Hence, by Proposition 6.1, R is Artinian.

Trivially (*b*) implies that R is Artinian. Now assume (*c*) and consider a prime ideal P of R. The domain R/P as a module over itself has its field of fractions as an injective envelope (see the example following Proposition 2.7). Since R/P is a self-injective ring, it follows that R/P is a field and P is a maximal ideal. Hence, by Theorem 4.6, R is Artinian. Thus, to prove the equivalence of (*a*), (*b*), (*c*), we can take R to be an Artinian ring and so a direct sum of local Artinian rings (Theorem 4.28), say

$$R = R_1 + R_2 + \ldots + R_n \quad \text{(d.s.).}$$

We must next see that it is sufficient to prove the equivalence of (*a*), (*b*), (*c*) when R is a local Artinian ring. We shall do this by showing that R satisfies (*a*) if and only if each ring R_i satisfies (*a*); and similarly for (*b*) and (*c*). Thus, before we proceed with the main part of the proof, we need to make some rather general remarks.

Write $$1_R = e_1 + e_2 + \ldots + e_n,$$

where $e_i \in R_i$ for $1 \leqslant i \leqslant n$. Then, for each i, e_i is the identity element of the ring R_i. Let L be an R-module. Then

$$L = e_1 L + e_2 L + \ldots + e_n L.$$

Further, this sum is direct. For suppose

$$e_1 x_1 + e_2 x_2 + \ldots + e_n x_n = e_1 x_1' + e_2 x_2' + \ldots + e_n x_n',$$

where $x_i, x_i' \in L$ for $1 \leqslant i \leqslant n$. Multiplying through by e_i, we have $e_i x_i = e_i x_i'$, because $e_i^2 = e_i$ and $e_i e_j \in R_i \cap R_j = 0$ whenever $i \neq j$. Also, $e_i L$ can be regarded as an R_i-module as well as an R-module, and its submodules are the same in both cases, because

$$(r_1 + r_2 + \ldots + r_n) e_i x = r_i e_i x,$$

where $r_j \in R_j$ for $1 \leqslant j \leqslant n$ and $x \in L$. Thus $R(e_i x) = R_i(e_i x)$, and the cyclic submodules of $e_i L$ are the same whether it is regarded as an R-module or as an R_i-module. It follows from these remarks that *if, for $1 \leqslant i \leqslant n$, every R_i-module is a direct sum of cyclic submodules, then every R-module is a direct sum of cyclic submodules.* Conversely, let $1 \leqslant i \leqslant n$, and consider an R_i-module L_i. We can regard L_i as an R-module: we write

$$(r_1 + r_2 + \ldots + r_n) x_i = r_i x_i,$$

where $r_j \in R_j$ for $1 \leqslant j \leqslant n$ and $x_i \in L_i$. Further, $Rx_i = R_i x_i$, so that the cyclic submodules of L_i are the same whether it is regarded as an R_i-module or as an R-module. Thus, *if every R-module is a direct sum of cyclic submodules, then, for* $1 \leqslant i \leqslant n$, *every R_i-module is a direct sum of cyclic submodules.* This deals with (*a*).

We now turn our attention to (*b*). Let I be an ideal of R. We can write
$$I = e_1 I + e_2 I + \ldots + e_n I \quad \text{(d.s.)},$$
and, for $1 \leqslant j \leqslant n$, $e_j I$ can be regarded as an ideal of the ring R_j. Suppose that, for $1 \leqslant j \leqslant n$, every ideal of R_j is principal. Then we can write $e_j I = R_j a_j$, where $a_j \in R_j$, and now
$$I = R(a_1 + a_2 + \ldots + a_n).$$
Thus every ideal of R is principal. Conversely, for $1 \leqslant j \leqslant n$, every ideal of R_j can be regarded as an ideal of R, and an R_j-ideal is principal as an R_j-ideal if and only if it is principal as an R-ideal. Thus, if every ideal of R is principal, then so is every ideal of R_j for $1 \leqslant j \leqslant n$.

It remains to consider (*c*). Let $1 \leqslant j \leqslant n$, and consider an ideal I_j of R_j. Let $\phi_j : R \to R_j$ be the projection mapping; this is a ring-homomorphism. Put $I = \phi_j^{-1}(I_j)$. Then I is an ideal of R and $R/I \approx R_j/I_j$, where this isomorphism is a ring-isomorphism. It follows that, *if R/I is a self-injective ring for every ideal I of R, then R_j/I_j is a self-injective ring for every ideal I_j of R_j.* Conversely, consider an ideal I of R. Now
$$R/I = ((R_1 + I)/I) + ((R_2 + I)/I) + \ldots + ((R_n + I)/I),$$
and this sum is easily seen to be direct. This expresses the ring R/I as a direct sum of rings. Further,
$$(R_j + I)/I \approx R_j/(I \cap R_j)$$
for $1 \leqslant j \leqslant n$, and this isomorphism is a ring-isomorphism. Thus, if we assume that, for $1 \leqslant j \leqslant n$, R_j/I_j is a self-injective ring for every ideal I_j of R_j, we have expressed R/I as a direct sum of self-injective rings. Thus, to deduce that R/I is a self-injective ring, *we need to show that a direct sum of self-injective rings is self-injective.* To do this, we shall simplify the notation. Suppose we have a direct sum
$$R = R_1 + R_2 + \ldots + R_n \quad \text{(d.s.)},$$

where the rings R_j are self-injective. Consider an R-module L which has R as a submodule. We have

$$L = e_1 L + e_2 L + \ldots + e_n L \quad \text{(d.s.)}.$$

For $1 \leqslant j \leqslant n$, $e_j L$ can be considered as an R_j-module, with R_j as a submodule. It follows from Theorem 2.15 that R_j is a direct summand of $e_j L$, as R_j-modules. Since the R-submodules and the R_j-submodules of $e_j L$ are the same, R_j is a direct summand of $e_j L$, as R-modules. Thus R is a direct summand of L. This shows that R is a self-injective ring.

To prove the equivalence of (a), (b) and (c), we can now add the assumption that R *is a local Artinian ring, with maximal ideal* M (*say*). The equivalence of (b) and (c) follows from Lemma 6.6 and Theorem 3.25 Corollary. Now assume (a) and consider a submodule N of $E(R/M)$. Then N is a direct sum of cyclic submodules. But the zero submodule of $E(R/M)$ is irreducible, so that N must actually be cyclic. It follows from Lemma 6.2 Corollary that the ideals of R are totally ordered whence, by Lemma 6.6, R is a principal ideal ring. This establishes (b).

It remains to assume (b) and (c) and deduce (a). We are also assuming that R is a local Artinian ring, so it follows from Lemma 6.6 that the ideals of R may be written as

$$R, \; Rp, \; Rp^2, \; \ldots, \; Rp^{s-1}, \; Rp^s = 0,$$

these being all distinct. Consider an R-module N. Let $\mathscr{A}$ be a maximal family of submodules of N of the form Rx, where $0:x = 0$, whose sum is direct. That such a family $\mathscr{A}$ exists follows from Proposition 1.7. Let F_0 be the sum of the members of $\mathscr{A}$. Then we may write
$$F_0 \approx \bigoplus_{i \in I_0} R.$$

Now R is a self-injective ring and is also Noetherian, so that F_0 is injective (Theorem 4.1). Hence we can write

$$N = F_0 + N_1 \quad \text{(d.s.)},$$

where N_1 is a submodule of N. Consider an element x_1 of N_1. Now $0:x_1 \neq 0$, for otherwise the family $\mathscr{A}$ could be enlarged by the submodule Rx_1. Let $r \in R$ be such that $rx_1 = 0$ and $r \neq 0$. We can write $r = up^l$, where u is a unit of R and $l < s$, so that $p^{s-1}x_1 = 0$.

This is true for all $x_1 \in N_1$, so that $\mathrm{Ann}_R N_1 \supseteq Rp^{s-1}$. If $s = 1$, this means that $N_1 = 0$. Suppose that $s > 1$. Put $\bar{R} = R/Rp^{s-1}$ and denote the natural image of p in $\bar{R}$ by $\bar{p}$. Now $\bar{R}$ as well as R is a local Artinian principal ideal ring, and its ideals are

$$\bar{R}, \ \bar{R}\bar{p}, \ \bar{R}\bar{p}^2, \ ..., \ \bar{R}\bar{p}^{s-1} = \bar{0}.$$

Further, N_1 may be regarded as an $\bar{R}$-module. It follows as above that we can write

$$N_1 = F_1 + N_2 \quad \text{(d.s.)},$$

where F_1 and N_2 are $\bar{R}$-submodules of N_1,

$$F_1 \approx \bigoplus_{i \in I_1} \bar{R}, \quad \mathrm{Ann}_{\bar{R}} N_2 \supseteq \bar{R}\bar{p}^{s-2}.$$

If we now regard N_1 as an R-module again, then F_1 and N_2 are R-submodules of N_1,

$$F_1 \approx \bigoplus_{i \in I_1} (R/Rp^{s-1}), \quad \mathrm{Ann}_R N_2 \supseteq Rp^{s-2}.$$

If $s = 2$, then $N_2 = 0$. If $s > 2$, we now regard N_2 as a module over the ring R/Rp^{s-2} and continue as before. In this way, we obtain

$$N = F_0 + F_1 + ... + F_{s-1} \quad \text{(d.s.)},$$

where
$$F_k \approx \bigoplus_{i \in I_k} R/Rp^{s-k}$$

for $0 \leqslant k \leqslant s - 1$. This establishes (a). $\square$

We wish to examine further the decomposition as a direct sum of cyclic submodules of a module over an Artinian principal ideal ring. In preparation for this, we make the following definition.

DEFINITION *A module is said to be 'uniserial' if it has only finitely many submodules and these are totally ordered.*

We note that every uniserial module is cyclic. For let A be a uniserial R-module, which we may suppose is non-zero, and let

$$A \supset A_1 \supset A_2 \supset ... \supset A_n = 0$$

be its chain of submodules. Let a be any element of A not in A_1. Then $Ra \nsubseteq A_1$, so that $Ra \supset A_1$ and $Ra = A$. Note also that A has A_1 as unique maximal submodule.

We also introduce a piece of notation. Let I be an ideal of R. If there is a non-negative integer n for which $I^n = I^{n+1}$,† then there is a least such integer, which we denote by $\lambda(I)$. If there is no such integer, we put $\lambda(I) = \infty$. We note that $\lambda(I) = 0$ if and only if $I = R$. If R is an Artinian ring, then $\lambda(I) < \infty$ for all I, because the decreasing sequence of ideals $I, I^2, I^3, \ldots$ must terminate. On the other hand, let R be a Noetherian domain and consider a proper ideal I of R. Suppose that n is a non-negative integer such that $I^n = I^{n+1}$. There is a maximal ideal M of R such that $I \subseteq M$, and $\bigcap_{k=1}^{\infty} M^k = 0$ by Proposition 4.23 Corollary 1.

Hence
$$I^n = \bigcap_{k=1}^{\infty} I^k \subseteq \bigcap_{k=1}^{\infty} M^k = 0,$$

whence $I = 0$. Thus, *when R is a Noetherian domain, $\lambda(I) = \infty$ for every proper non-zero ideal I of R.*

In our next lemma, we shall describe the uniserial modules in the situation which will arise in the sequel.

LEMMA 6.8 *Let R be a ring such that, for every maximal ideal M of R and every positive integer k, R/M^k is an Artinian principal ideal ring. Let A be an R-module and let n be a positive integer. Then the following statements are equivalent:*

(a) *A is a uniserial module with n proper submodules;*

(b) *there is a maximal ideal P of R such that $\lambda(P) \geqslant n$ and $A \approx R/P^n$.*

Proof. Let P be a maximal ideal of R such that $\lambda(P) \geqslant n$. By hypothesis, R/P^n is an Artinian principal ideal ring, and it is local since P is the only maximal ideal of R which contains P^n. It follows from Lemma 6.6 that the ideals of R/P^n are all of the form P^r/P^n for $0 \leqslant r \leqslant n$. Hence the submodules of R/P^n, considered now as an R-module, are all of the form P^r/P^n for $0 \leqslant r \leqslant n$; and they are all different because $\lambda(P) \geqslant n$. Hence (b) implies (a).

To prove that (a) implies (b), we use induction on n. Assume (a) with $n = 1$. Then A is simple, so there is a maximal ideal P of R such that $A \approx R/P$; certainly $\lambda(P) \geqslant 1$. Now let $n > 1$ and assume that (a) implies (b) for $n-1$. Suppose that A is

† We define $I^0 = R$.

uniserial with n proper submodules. Since A is cyclic, there is an ideal I of R such that $A \approx R/I$. Let

$$A \supset A_1 \supset A_2 \supset \ldots \supset A_{n-1} \supset A_n = 0$$

be the chain of submodules of A. Now A/A_{n-1} is a uniserial R-module with $n-1$ proper submodules, so there is a maximal ideal P of R such that $\lambda(P) \geqslant n-1$ and

$$A/A_{n-1} \approx R/P^{n-1}.$$

Also, A_{n-1} is simple, so there is a maximal ideal P' of R such that

$$A_{n-1} \approx R/P'.$$

Now $\qquad I = \mathrm{Ann}_R A \subseteq \mathrm{Ann}_R A_{n-1} = P'$

and $\qquad I = \mathrm{Ann}_R A \subseteq \mathrm{Ann}_R(A/A_{n-1}) = P^{n-1} \subseteq P.$

But R/I is a uniserial R-module, so that I is contained in a unique maximal ideal of R. It follows that $P' = P$. Further, $P^{n-1}(A/A_{n-1}) = 0$, so that $P^n A \subseteq P A_{n-1} = 0$. It follows that $P^n \subseteq I \subseteq P^{n-1}$. By hypothesis, R/P^n is an Artinian principal ideal ring; it is also local, with maximal ideal P/P^n. It follows from Lemma 6.6 that all the ideals of R/P^n are of the form P^r/P^n for $0 \leqslant r \leqslant n$. Hence either $I = P^{n-1}$ or $I = P^n$. But I and P^{n-1} cannot be equal, because R/I and R/P^{n-1} have a different number of submodules. Hence $I = P^n$, $A \approx R/P^n$ and $\lambda(P) \geqslant n.\square$

Now let R be an Artinian principal ideal ring and let M be an R-module. Then R is a direct sum of local Artinian principal ideal rings, say

$$R = R_1 + R_2 + \ldots + R_t \quad \text{(d.s.)}.$$

Let $1 = e_1 + e_2 + \ldots + e_t$, where $e_i \in R_i$ for $1 \leqslant i \leqslant t$. Then e_i is the identity element of the ring R_i. Also,

$$M = e_1 M + e_2 M + \ldots + e_t M \quad \text{(d.s.)}.$$

Now $e_k M$ may be regarded as a module over the local Artinian principal ideal ring R_k, and as such it is a direct sum of cyclic submodules. By Lemma 6.6, R_k is uniserial as a module over itself. We denote by μ_k the number of ideals of R_k. It follows that a cyclic R_k-module is uniserial, having at most μ_k submodules.

Thus $e_k M$ is a direct sum of uniserial R_k-submodules, each having at most μ_k submodules. Since the R-submodules and the R_k-submodules of $e_k M$ are one and the same, it follows that $e_k M$ is a direct sum of uniserial R-modules, each having at most μ_k submodules. Now put $\mu = \max_{1 \leqslant k \leqslant t} \mu_k$. Then M *is a direct sum of uniserial submodules, each summand having at most μ submodules.* We shall use this information in its present form later on. But we are in a position to use Lemma 6.8. This gives us the existence part of the next theorem.

THEOREM 6.9 *Let R be an Artinian principal ideal ring and let M be an R-module. Then there is a unique family $\{P_i, n_i\}_{i \in I}$ such that:*

(i) *the P_i are maximal ideals of R;*

(ii) *the n_i are positive integers with $n_i \leqslant \lambda(P_i)$ for every i;*

(iii) $M \approx \underset{i \in I}{\oplus} R/P_i^{n_i}.$

Proof. It remains to prove uniqueness. Consider a maximal ideal P of R and a positive integer n. We may regard R/P^n as a module over either of the rings R and R/P^n, and its endomorphisms are the same in each case. But the ring of endomorphisms of R/P^n as a module over itself is isomorphic to R/P^n, which is a local ring having P/P^n as its maximal ideal. Theorem 3.13 now gives that the family of R-modules $\{R/P_i^{n_i}\}_{i \in I}$ is unique up to isomorphism.

Suppose now that $\qquad R/P^m \approx R/Q^n,$

where P, Q are maximal ideals of R and $0 < m \leqslant \lambda(P)$, $0 < n \leqslant \lambda(Q)$. If we compare annihilators, we see that $P^m = Q^n$, whence $P^m \subseteq Q, P \subseteq Q$ and finally $P = Q$. We now have $P^m = P^n$. Since $0 < m, n \leqslant \lambda(P)$, this means that $m = n$. Thus the family $\{P_i, n_i\}_{i \in I}$ is actually unique. $\square$

LEMMA 6.10 *An Artinian ring has only finitely many maximal ideals.*

Proof. Let R be an Artinian ring and denote by Ω the set of all ideals which are the intersection of a finite number of maximal ideals of R. Then Ω is non-empty (by convention, the intersection of no maximal ideals of R is R itself) and so has a minimal member, say $P_1 \cap P_2 \cap \ldots \cap P_n$ $(n \geqslant 0)$, where the P_i are maximal ideals of R. Let P be a maximal ideal of R. Now

$$P_1 \cap P_2 \cap \ldots \cap P_n \cap P \subseteq P_1 \cap P_2 \cap \ldots \cap P_n,$$

so that $P_1 \cap P_2 \cap \ldots \cap P_n \cap P = P_1 \cap P_2 \cap \ldots \cap P_n$

and $P \supseteq P_1 \cap P_2 \cap \ldots \cap P_n \supseteq P_1 P_2 \ldots P_n.$†

Since the P_i are prime, it follows that $P \supseteq P_i$ for some i, whence $P = P_i$. It follows that $P_1, P_2, \ldots, P_n$ are all the maximal ideals of R.☐

We shall be interested in those commutative domains which satisfy the equivalent conditions of Theorem 6.7 residually, i.e. those commutative domains R such that, for every non-zero ideal I of R, every module over the residue class ring R/I is a direct sum of cyclic submodules. *This is equivalent to the condition that every R-module with a non-zero annihilator is a direct sum of cyclic submodules.* We shall see in Theorem 6.14 that this is equivalent to the condition that R is a Dedekind domain.

THEOREM 6.11   *Let R be a commutative domain with the property that R/I is an Artinian principal ideal ring for every non-zero ideal I of R. Let M be an R-module with a non-zero annihilator. Then there is a unique family $\{P_i, n_i\}_{i \in I'}$ such that:*

(i)   *the P_i are non-zero maximal ideals of R and there are only finitely many distinct ones;*

(ii) *$\{n_i\}_{i \in I'}$ is a bounded family of positive integers;*

(iii) *$M \approx \underset{i \in I'}{\oplus} R/P_i^{n_i}$.*

REMARK   Consider the R-module $\underset{i \in I'}{\oplus} R/P_i^{n_i}$, where R is a domain and conditions (i) and (ii) of the theorem are satisfied. Let $P_{i_1}, P_{i_2}, \ldots, P_{i_t}$ denote the distinct maximal ideals among the P_i and let n be an upper bound for the family $\{n_i\}_{i \in I'}$. Then the annihilator of this module contains $P_{i_1}^n \cap P_{i_2}^n \cap \ldots \cap P_{i_t}^n$, which in turn contains $P_{i_1}^n P_{i_2}^n \ldots P_{i_t}^n$. This ideal is non-zero because R is a domain. Thus Theorem 6.11 gives a complete description of the R-modules with non-zero annihilator when R satisfies the conditions of the theorem.

† If $n = 0$, this product is taken to be R.

Proof of Theorem 6.11. Put $\bar{R} = R/\mathrm{Ann}_R M$. Then $\bar{R}$ is an Artinian principal ideal ring, and M can be regarded as an $\bar{R}$-module. As such, M is a direct sum of uniserial submodules such that there is an upper bound n (say) on the number of proper submodules that any of the uniserial summands can possess. Now the R-submodules of M and its $\bar{R}$-submodules are one and the same, so that, as an R-module,

$$M = \sum_{i \in I'} M_i \quad \text{(d.s.)},$$

where the M_i are non-zero uniserial submodules of M with at most n proper submodules. Now the domain R satisfies the requirements of Lemma 6.8† so that, for each i, we can write $M_i \approx R/P_i^{n_i}$, where n_i is the number of proper submodules of M_i and P_i is a maximal ideal of R. Also,

$$0 \neq \mathrm{Ann}_R M \subseteq P_i^{n_i} \subseteq P_i$$

for every i. The ideals $P_i/\mathrm{Ann}_R M$ of the ring $R/\mathrm{Ann}_R M$ are maximal and this ring is Artinian, so by Lemma 6.10 there are only finitely many distinct ideals $P_i/\mathrm{Ann}_R M$. It follows that there are only finitely many distinct ideals P_i. This establishes the existence of a family $\{P_i, n_i\}_{i \in I'}$ with the required properties.

The proof of the uniqueness of the family $\{P_i, n_i\}_{i \in I'}$ is the same as in Theorem 6.9. There is one point, however, that needs attention. To apply the proof given for Theorem 6.9, we need to know that $n_i \leqslant \lambda(P_i)$ for each i. Now the ring R/I is Artinian and so Noetherian for every non-zero ideal I of R (see Theorem 3.25 Corollary). *But this means that R itself is Noetherian.* For consider an ascending chain $I_1 \subseteq I_2 \subseteq I_3 \subseteq \cdots$

of non-zero ideals of R. This gives rise to the ascending chain

$$0 \subseteq I_2/I_1 \subseteq I_3/I_1 \subseteq \cdots$$

of ideals of R/I_1, which must terminate because R/I_1 is a Noetherian ring. Hence the original chain of ideals of R must terminate. Thus R is a Noetherian domain, and we have already remarked that, in this situation, $\lambda(I) = \infty$ for every proper non-zero ideal I of R. Hence the condition $n_i \leqslant \lambda(P_i)$ is automatically satisfied for every i. □

---

† Distinguish here between the cases when R is and is not a field.

6.2 Characterizations of Dedekind domains

In our next theorem, we shall prove that the commutative domains which satisfy the equivalent conditions of Theorem 6.7 residually are precisely the Dedekind domains. We need two lemmas, both of an elementary nature.

LEMMA 6.12   *Let* $P_1, P_2, ..., P_k$ $(k \geqslant 1)$ *be distinct maximal ideals of R and let* $n_1, n_2, ..., n_k$ *be positive integers. Then*

$$P_1^{n_1} \cap P_2^{n_2} \cap ... \cap P_k^{n_k} = P_1^{n_1} P_2^{n_2} ... P_k^{n_k}.$$

Proof. We use induction on k, the result being obvious when $k = 1$. Let $k > 1$, and assume that

$$P_1^{n_1} \cap P_2^{n_2} \cap ... \cap P_{k-1}^{n_{k-1}} = P_1^{n_1} P_2^{n_2} ... P_{k-1}^{n_{k-1}}.$$

Suppose that $(P_1^{n_1} P_2^{n_2} ... P_{k-1}^{n_{k-1}}) + P_k^{n_k}$

is a proper ideal of R. Then there is a maximal ideal P such that

$$P_1^{n_1} P_2^{n_2} ... P_{k-1}^{n_{k-1}} \subseteq P, \quad P_k^{n_k} \subseteq P,$$

whence $P_i \subseteq P$ for some i, $1 \leqslant i \leqslant k-1$, and $P_k \subseteq P$. This means that $P_i = P = P_k$, which is not so. Thus

$$(P_1^{n_1} P_2^{n_2} ... P_{k-1}^{n_{k-1}}) + P_k^{n_k} = R,$$

so there exist $\alpha \in P_1^{n_1} P_2^{n_2} ... P_{k-1}^{n_{k-1}}$ and $\beta \in P_k^{n_k}$ such that $\alpha + \beta = 1$. Let $\gamma \in P_1^{n_1} \cap P_2^{n_2} \cap ... \cap P_k^{n_k}$. Now $\gamma = \alpha\gamma + \beta\gamma$. Further,

$$\alpha\gamma \in P_1^{n_1} P_2^{n_2} ... P_k^{n_k}$$

and

$$\beta\gamma \in (P_1^{n_1} \cap P_2^{n_2} \cap ... \cap P_{k-1}^{n_{k-1}}) P_k^{n_k} = P_1^{n_1} P_2^{n_2} ... P_{k-1}^{n_{k-1}} P_k^{n_k}.$$

Hence $\gamma \in P_1^{n_1} P_2^{n_2} ... P_k^{n_k}$, which shows that

$$P_1^{n_1} \cap P_2^{n_2} \cap ... \cap P_k^{n_k} \subseteq P_1^{n_1} P_2^{n_2} ... P_k^{n_k}.$$

But the opposite inclusion is obvious.☐

For the next lemma, R will be a commutative domain. We recall from Section 2.2 that a fractional ideal F of R is invertible if there is a fractional ideal F^{-1} such that $FF^{-1} = R$.

We adopt the convention that the product of no ideals of a ring shall be the ring itself.

LEMMA 6.13 *There is at most one way of expressing an ideal of R as a product of invertible prime ideals, apart from the order of the terms.*

Proof. Suppose that

$$P_1 P_2 \ldots P_r = Q_1 Q_2 \ldots Q_s,$$

where the P_i and Q_j are invertible prime ideals of R. We use induction on $\min(r, s)$. If $\min(r, s) = 0$, the result is obvious. We now suppose that $\min(r, s) > 0$. We may suppose that P_r is minimal among the P_i and the Q_j. Now

$$Q_1 Q_2 \ldots Q_s = P_1 P_2 \ldots P_r \subseteq P_r,$$

so that $Q_j \subseteq P_r$ for some j, say $Q_s \subseteq P_r$. Since P_r is minimal, this means that $Q_s = P_r$. If we now multiply by P_r^{-1}, we have

$$P_1 P_2 \ldots P_{r-1} = Q_1 Q_2 \ldots Q_{s-1},$$

and the inductive hypothesis can be used. □

THEOREM 6.14 *Let R be a commutative domain. Then the following statements are equivalent:*

(a) every R-module with a non-zero annihilator is a direct sum of cyclic submodules;

(b) for every non-zero ideal I of R, R/I is an Artinian principal ideal ring;

(c) R is Noetherian and, for every non-zero ideal I of R, the ring R/I is self-injective;

(d) every ideal of R is a product of prime ideals;

(e) R is a Dedekind domain.

Proof. We demonstrated in the proof of Theorem 6.11 that, if the ring R/I is Noetherian for every non-zero ideal I of R, then R itself is Noetherian. Also, the statement that, for every non-zero ideal I of R, every residue class ring of R/I is self-injective is equivalent to the statement that, for every non-zero ideal A of R, the ring R/A is self-injective.† The equivalence of (a), (b), (c) now follows from Theorem 6.7.

† We make use here of the isomorphism $(R/I)/(A/I) \approx R/A$ given in Proposition 1.10 Corollary 2, where I and A are ideals of R such that $A \supseteq I$. Although this isomorphism is originally an R-isomorphism, it is easily seen to be an isomorphism of rings.

Now assume (*b*), and consider an ideal A of R. We shall show that A can be expressed as a product of prime ideals, and for this we may suppose that $A \neq 0$. Consider the R-module R/A. This is finitely generated, and has the non-zero annihilator A. It follows from Theorem 6.11 that there are maximal ideals $P_1, P_2, ..., P_k$ and positive integers $n_1, n_2, ..., n_k$ such that

$$R/A \approx (R/P_1^{n_1}) \oplus (R/P_2^{n_2}) \oplus ... \oplus (R/P_k^{n_k}).$$

If we now compare annihilators, we have

$$A = P_1^{n_1} \cap P_2^{n_2} \cap ... \cap P_k^{n_k}.$$

By omitting any redundant terms, we may suppose in this intersection that $P_1, P_2, ..., P_k$ are all different. Now Lemma 6.12 gives that
$$A = P_1^{n_1} P_2^{n_2} ... P_k^{n_k}.$$
This establishes (*d*).

We now assume (*d*) and prove that R is a Dedekind domain. We first show that *every invertible prime ideal is maximal*. Let P be an invertible prime ideal of R and consider any element $a \in R$, $a \notin P$. We shall show that $P + Ra = R$. Suppose that $P + Ra \neq R$. By (*d*), we can write

$$P + Ra = Q_1 Q_2 ... Q_s,$$

$$P + Ra^2 = Q_1' Q_2' ... Q_t',$$

where the Q_i and the Q_j' are prime ideals of R and $s, t \geqslant 1$. We now pass to the residue class ring $R/P = R^*$ (say), and use a star to denote the image of an element or ideal of R under the natural mapping $R \to R/P$. Note that R^* is an integral domain because P is a prime ideal of R. Then

$$R^* a^* = Q_1^* Q_2^* ... Q_s^*,$$

$$R^* a^{*2} = Q_1'^* Q_2'^* ... Q_t'^*.$$

Now $R^* a^*$ and $R^* a^{*2}$ are non-zero principal ideals of R^* and so are invertible. Hence the Q_i^* and $Q_j'^*$ are invertible ideals of R^*; they are also prime because the ideals Q_i and Q_j' are prime ideals of R. Further,

$$Q_1^{*2} Q_2^{*2} ... Q_s^{*2} = Q_1'^* Q_2'^* ... Q_t'^*.$$

It follows from Lemma 6.13 that $t = 2s$ and that we can write

$$Q_1'^* = Q_2'^* = Q_1^*, \quad Q_3'^* = Q_4'^* = Q_2^* \quad \text{etc.}$$

Thus $\qquad Q_1' = Q_2' = Q_1, \quad Q_3' = Q_4' = Q_2 \quad$ etc.

and $\quad P + Ra^2 = (P + Ra)^2 = P^2 + RaP + Ra^2 \subseteq P^2 + Ra.$

Hence $P \subseteq P^2 + Ra$. Consider an element α of P. We can write $\alpha = \beta + ra$, where $\beta \in P^2$ and $r \in R$. Then $ra = \alpha - \beta \in P$ and $a \notin P$, so $r \in P$. Hence $P \subseteq P^2 + aP$. But obviously $P^2 + aP \subseteq P$, so that

$$P = P^2 + aP = P(P + Ra).$$

But P is invertible, so that $P + Ra = R$, which is a contradiction.

We now show that *every non-zero prime ideal is invertible*. Let P' be a non-zero prime ideal of R and let x be a non-zero element of P'. By (d), we can write

$$Rx = P_1 P_2 \dots P_r,$$

where the P_i are prime ideals of R. Now Rx is invertible, so each P_i is invertible. It follows from the above that each P_i is maximal. Now

$$P' \supseteq Rx = P_1 P_2 \dots P_r,$$

so that $P' \supseteq P_i$ for some i, whence $P' = P_i$. Thus P' is invertible.

We can now deduce that R is a Dedekind domain. By Lemma 2.9, we must show that every non-zero integral ideal of R is invertible. But, by (d), a non-zero integral ideal is a product of non-zero prime ideals, and each of these prime ideals is invertible.

To complete the proof, we assume that R is a Dedekind domain and prove that R satisfies (c). By Proposition 2.10, R has the property that every divisible R-module is injective, so that direct sums of injective modules are injective and R is Noetherian (Theorem 4.1). Now consider a non-zero ideal I of R, and denote by K the quotient field of R. Then K may be regarded as an R-module; as such it is divisible. It follows that K/I is a divisible R-module, and so injective. Hence, by Proposition 2.27, $0 :_{K/I} I$ is an injective (R/I)-module. We write $0 :_{K/I} I = A/I$, where A is a submodule of K containing I. Since $I(R/I) = 0$, we have $R \subseteq A$.

Also, $I(A/I) = 0$, so that $IA \subseteq I$. But I is invertible because R is a Dedekind domain, so that

$$A = I^{-1}IA \subseteq I^{-1}I = R.$$

Thus $A = R$ and R/I is a self-injective ring. $\square$

In view of Theorem 6.14, *Theorem 6.11 describes the structure of a module with a non-zero annihilator over a Dedekind domain.*

6.3   Finitely generated modules over Dedekind domains

We shall now go on to describe the structure of a finitely generated module over a Dedekind domain. From this result, we shall be able to say which commutative Noetherian domains R have the property that every finitely generated R-module is a direct sum of cyclic submodules.

We first prove an elementary lemma.

LEMMA 6.15   *Let R be a commutative domain with quotient field K and let I be a finitely generated submodule of K, where K is considered as an R-module. Then I is a fractional ideal of R.*

Proof. We put

$$I = R(a_1/b_1) + R(a_2/b_2) + \ldots + R(a_s/b_s),$$

where the a_i and b_i are elements of R with the b_i non-zero. Then $(b_1 b_2 \ldots b_s)I \subseteq R$ and $b_1 b_2 \ldots b_s \neq 0$, so that I is a fractional ideal of R. $\square$

THEOREM 6.16   *Let R be a Dedekind domain and let M be a finitely generated R-module. Then*

$$M \approx (R/P_1^{n_1}) \oplus (R/P_2^{n_2}) \oplus \ldots \oplus (R/P_k^{n_k}) \oplus I_1 \oplus I_2 \oplus \ldots \oplus I_l,$$

where

  (i) k, l *are non-negative integers,*
  (ii) $P_1, P_2, \ldots, P_k$ *are non-zero maximal ideals of R,*
  (iii) $n_1, n_2, \ldots, n_k$ *are positive integers,*
  (iv) $I_1, I_2, \ldots, I_l$ *are non-zero fractional ideals of R.*

Proof. We write $M = Rm_1 + Rm_2 + \ldots + Rm_t$. Then

$$\mathrm{Ann}_R M = (0{:}m_1) \cap \ldots \cap (0{:}m_t) \supseteq (0{:}m_1)(0{:}m_2) \ldots (0{:}m_t).$$

First suppose that $\mathrm{Ann}_R M = 0$. Since R is a domain, this means that there is an element m of M such that $0:m = 0$. With such an element m, we consider the diagram Fig. 6.2, where the mapping θ is given by $\theta(r) = rm$ for $r \in R$, K is the quotient field of R considered as an R-module and $R \to K$ is the inclusion mapping. Since $0:m = 0$, θ is a mono-morphism. Also, K is injective; it is

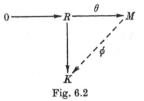

Fig. 6.2

actually an injective envelope of R. We can thus complete the diagram by means of an R-homomorphism $\phi: M \to K$. Note that $\phi(m) = 1$, so that ϕ is not the zero mapping. We put $I_1 = \phi(M)$. Then I_1 is a finitely generated submodule of K, and so by Lemma 6.15 is a fractional ideal of R. Since $I_1 \neq 0$, it has an inverse I_1^{-1}, so that $I_1 I_1^{-1} = R$. Thus there exist elements $\alpha_i \in I_1$ and $\beta_i \in I_1^{-1}$ $(1 \leqslant i \leqslant s)$ such that $\sum_{i=1}^{s} \alpha_i \beta_i = 1$. There now exist elements $m_i' \in M$ such that $\phi(m_i') = \alpha_i$. We define an R-homomorphism $\psi_1: I_1 \to M$ by

$$\psi_1(x) = \sum_{i=1}^{s} (x\beta_i) m_i', \quad \text{where} \quad x \in I_1.$$

Then, for $x \in I_1$, $\qquad \phi\psi_1(x) = \sum_{i=1}^{s} x\beta_i\alpha_i = x,$

so that $\phi\psi_1$ is the identity mapping of I_1. It follows that ψ_1 is a monomorphism. Also, for $y \in M$,

$$y = \psi_1\phi(y) + (y - \psi_1\phi(y)) \in \psi_1(I_1) + \mathrm{Ker}\,\phi,$$

so that $M = \psi_1(I_1) + \mathrm{Ker}\,\phi$. Moreover, this sum is direct since $\psi_1(I_1) \cap \mathrm{Ker}\,\phi = 0$. Put $M_1 = \mathrm{Ker}\,\phi$. Then

$$M = \psi_1(I_1) + M_1 \quad \text{(d.s.)}.$$

Now R is a Noetherian domain, so that M is Noetherian and M_1 is finitely generated. If M_1 has zero annihilator, it follows from the above that there is a non-zero fractional ideal I_2 of R, a mono-morphism $\psi_2: I_2 \to M$ and a submodule M_2 of M such that

$$M = \psi_1(I_1) + \psi_2(I_2) + M_2 \quad \text{(d.s.)}.$$

We continue in this way. Since M is Noetherian, this process must terminate. When this happens, we have non-zero fractional ideals $I_1, I_2, ..., I_l$ and an R-module N with a non-zero annihilator such that

$$M \approx I_1 \oplus I_2 \oplus ... \oplus I_l \oplus N.$$

Now N is finitely generated, so that, by Theorem 6.11,

$$N \approx (R/P_1^{n_1}) \oplus (R/P_2^{n_2}) \oplus ... \oplus (R/P_k^{n_k}),$$

where the P_i are non-zero maximal ideals of R and the n_i are positive integers. □

REMARK In Theorem 6.16, it can be arranged that the fractional ideals $I_1, I_2, ..., I_l$ are integral ideals of R. This is because, for each u ($1 \leqslant u \leqslant l$), there is a non-zero element γ_u of R such that $\gamma_u I_u \subseteq R$. Then $\gamma_u I_u$ is a non-zero integral ideal of R and the mapping $I_u \to \gamma_u I_u$ given by multiplication by γ_u is an isomorphism.

COROLLARY *Let R be a commutative Noetherian domain. Then the following statements are equivalent:*

(a) every finitely generated R-module is a direct sum of cyclic submodules;

(b) R is a principal ideal domain.

Proof. Since a principal ideal domain is Dedekind, it follows from Theorem 6.16 and the above remark that (b) implies (a). Next, if A_1 and A_2 are non-zero integral ideals of R, then $A_1 \cap A_2 \supseteq A_1 A_2$, which is non-zero since R is a domain. This means that the non-zero integral ideals of R are indecomposable. Thus, if R satisfies (a), then every finitely generated integral ideal of R, and so every integral ideal of R, must be cyclic, i.e. R is a principal ideal domain. □

In Theorem 6.16, we cannot say that, for a given M, the fractional ideals $I_1, I_2, ..., I_l$ are uniquely determined up to isomorphism. We shall thus modify our decomposition of a finitely generated module over a Dedekind domain to obtain a decomposition which is essentially unique.

DEFINITION *Let R be a domain with field of fractions K and let I, J be fractional ideals of R. We say that I and J 'belong to the same ideal class' if there is a non-zero element α of K such that $I = \alpha J$.*

This gives an equivalence relation on the fractional ideals of R, and so partitions the set of fractional ideals into equivalence classes, the *ideal classes* of R. Two fractional ideals which belong to the same ideal class are isomorphic; for if I is a fractional ideal of R and α is a non-zero element of K, then the mapping $\phi: I \to \alpha I$ given by $\phi(x) = \alpha x$, where $x \in I$, is an R-isomorphism. Also, if for $1 \leqslant r \leqslant k$ the fractional ideals I_r, J_r belong to the same ideal class, then $I_1 I_2 \ldots I_k$ and $J_1 J_2 \ldots J_k$ belong to the same ideal class.

LEMMA 6.17   *Let R be a domain with field of fractions K and let I be a non-zero fractional ideal of R. Then K is an injective envelope of I.*

Proof. We know from the example following Proposition 2.7 that K is injective as an R-module. Now consider a non-zero element $x \in K$. There exists $r \in R$ such that rx is a non-zero element of R. Let $\beta \in R$ be a non-zero element of I. Then βrx is a non-zero element of I. This shows that K is an essential extension of I. Thus $K = E(I)$. □

LEMMA 6.18   *Let R be a Dedekind domain with field of fractions K and suppose that*

$$I_1 \oplus I_2 \oplus \ldots \oplus I_m \approx J_1 \oplus J_2 \oplus \ldots \oplus J_n,$$

where the I_r and the J_s are non-zero fractional ideals of R. Then $m = n$ and $I_1 I_2 \ldots I_m, J_1 J_2 \ldots J_n$ belong to the same ideal class.

Proof. Let

$$\phi: I_1 \oplus I_2 \oplus \ldots \oplus I_m \to J_1 \oplus J_2 \oplus \ldots \oplus J_n$$

be an isomorphism. This isomorphism extends to an isomorphism between the respective injective envelopes, so by Proposition 2.23 and Lemma 6.17 ϕ extends to an R-isomorphism

$$\bar{\phi}: K \oplus K \oplus \ldots \oplus K \to K \oplus K \oplus \ldots \oplus K,$$

where there are m summands on the left hand side and n on the right hand side. Now $K = E(R)$, so by Proposition 2.28 Corollary 1 K is an indecomposable injective R-module. It follows from Theorem 3.13 Corollary that $m = n$.

We may regard $K \oplus K \oplus \ldots \oplus K$ in the obvious way as a vector space over K, when $\overline{\phi}$ becomes a K-linear mapping. For if $a, b \in R$, $b \neq 0$ and $\xi \in K \oplus K \oplus \ldots \oplus K$, then

$$b\overline{\phi}(ab^{-1}\xi) = \overline{\phi}(a\xi) = a\overline{\phi}(\xi),$$

whence

$$\overline{\phi}(ab^{-1}\xi) = ab^{-1}\overline{\phi}(\xi).$$

For $1 \leqslant r \leqslant n$, we put

$$\overline{\phi}(0, \ldots, 1, \ldots, 0) = (\alpha_{r1}, \ldots, \alpha_{rn}),$$

where 1 occurs as the rth component, the remaining components being zero. For $\gamma_r \in I_r$ we have

$$\phi(0, \ldots, \gamma_r, \ldots, 0) = \overline{\phi}(0, \ldots, \gamma_r, \ldots, 0) = (\gamma_r \alpha_{r1}, \ldots, \gamma_r \alpha_{rn}),$$

which shows that $\alpha_{rs}I_r \subseteq J_s$ for $1 \leqslant r, s \leqslant n$. Let $s_1, s_2, \ldots, s_n$ be any rearrangement of $1, 2, \ldots, n$. Then

$$J_1 J_2 \ldots J_n = J_{s_1} J_{s_2} \ldots J_{s_n} \supseteq \alpha_{1s_1} \alpha_{2s_2} \ldots \alpha_{ns_n} I_1 I_2 \ldots I_n,$$

whence

$$J_1 J_2 \ldots J_n \supseteq \Delta I_1 I_2 \ldots I_n,$$

where

$$\Delta = \det \|\alpha_{rs}\|.$$

For $1 \leqslant r \leqslant n$, we put

$$\overline{\phi}^{-1}(0, \ldots, 1, \ldots, 0) = (\beta_{r1}, \ldots, \beta_{rn}),$$

where 1 occurs as the rth component, the remaining components being zero. Now $\overline{\phi}^{-1}$ extends ϕ^{-1}, so for $\delta_r \in J_r$ we have

$$\phi^{-1}(0, \ldots, \delta_r, \ldots, 0) = \overline{\phi}^{-1}(0, \ldots, \delta_r, \ldots, 0) = (\delta_r \beta_{r1}, \ldots, \delta_r \beta_{rn}).$$

Thus, just as above, we have

$$I_1 I_2 \ldots I_n \supseteq \Delta' J_1 J_2 \ldots J_n,$$

where $\Delta' = \det \|\beta_{rs}\|$. But $\|\alpha_{rs}\|$ and $\|\beta_{rs}\|$ are inverse matrices, so that $\Delta' = \Delta^{-1}$ and

$$\Delta I_1 I_2 \ldots I_n \supseteq J_1 J_2 \ldots J_n.$$

This shows that

$$J_1 J_2 \ldots J_n = \Delta I_1 I_2 \ldots I_n,$$

so that $I_1 I_2 \ldots I_n$ and $J_1 J_2 \ldots J_n$ are in the same ideal class. $\square$

LEMMA 6.19 *Let R be a Dedekind domain and let*

$$I_1, I_2, \ldots, I_l \quad (l \geqslant 1)$$

be non-zero fractional ideals of R. Then

$$I_1 \oplus I_2 \oplus \ldots \oplus I_l \approx R \oplus R \oplus \ldots \oplus R \oplus (I_1 I_2 \ldots I_l),$$

where there are $l-1$ summands R on the right hand side.

Proof. The case $l = 1$ is trivial; the general case will follow by induction from the case $l = 2$. Accordingly, we consider non-zero fractional ideals I_1, I_2. Since R is a Dedekind domain, we can invert I_1 and I_2. Thus there is a non-zero element γ of R such that $\gamma I_2^{-1} \subseteq R$. Let α be a non-zero element of I_1^{-1}. Then

$$R \supseteq \gamma I_2^{-1} \supseteq (\alpha I_1)(\gamma I_2^{-1}) \neq 0,$$

so that $R/(\alpha I_1)(\gamma I_2^{-1})$ is a principal ideal ring (see Theorem 6.14). Hence there is an element $\beta \in R$ such that

$$\gamma I_2^{-1}/(\alpha I_1)(\gamma I_2^{-1}) = (\beta R + (\alpha I_1)(\gamma I_2^{-1}))/(\alpha I_1)(\gamma I_2^{-1}),$$

whence $\qquad \gamma I_2^{-1} = \beta R + (\alpha I_1)(\gamma I_2^{-1}),$

$$R = \beta \gamma^{-1} I_2 + \alpha I_1.$$

If $\beta = 0$, then $R = \alpha I_1$ and

$$I_1 \oplus I_2 \approx R \oplus I_2 = R \oplus R I_2 \approx R \oplus I_1 I_2,$$

as required. Suppose now that $\beta \neq 0$. Then αI_1 resp. $\beta \gamma^{-1} I_2$ belong to the same ideal class as I_1 resp. I_2. Thus, in proving that

$$I_1 \oplus I_2 \approx R \oplus I_1 I_2,$$

we may suppose that $I_1 + I_2 = R$. Hence there exist $\alpha_1 \in I_1$ and $\alpha_2 \in I_2$ such that $\alpha_1 + \alpha_2 = 1$. Also $I_1 I_2 \subseteq I_1$ and $I_1 I_2 \subseteq I_2$. We define the R-homomorphisms

$$\theta: I_1 \oplus I_2 \to R \oplus I_1 I_2,$$

$$\theta': R \oplus I_1 I_2 \to I_1 \oplus I_2$$

by $\qquad \theta(\beta_1, \beta_2) = (\beta_1 + \beta_2, \alpha_1 \beta_2 - \alpha_2 \beta_1),$

$$\theta'(\gamma, \delta) = (\alpha_1 \gamma - \delta, \alpha_2 \gamma + \delta),$$

where $\beta_1 \in I_1$, $\beta_2 \in I_2$, $\gamma \in R$, $\delta \in I_1 I_2$. Then $\theta\theta'$ and $\theta'\theta$ are both identity mappings and so are isomorphisms. $\square$

Let M be a module over a commutative domain R. We denote by $T(M)$ the set of all torsion elements of M, i.e. the set of all elements m of M for which there exists a non-zero element r of R such that $rm = 0$. Then $T(M)$ is a submodule of M, and is called the *torsion submodule* of M. If $T(M) = M$, then M is said to be a *torsion module*; and M is *torsion-free* precisely when $T(M) = 0$. An isomorphism between two R-modules restricts to an isomorphism between their torsion submodules.

Now let M be a finitely generated R-module, say

$$M = Rm_1 + Rm_2 + \ldots + Rm_t.$$

Then $\qquad \operatorname{Ann}_R M = (0:m_1) \cap (0:m_2) \cap \ldots \cap (0:m_t)$

$$\supseteq (0:m_1)(0:m_2) \ldots (0:m_t),$$

from which it follows that M is a torsion module if and only if $\operatorname{Ann}_R M \neq 0$. Thus Theorem 6.11 tells us the structure of a finitely generated torsion module M over a Dedekind domain R; it is isomorphic to

$$(R/P_1^{n_1}) \oplus (R/P_2^{n_2}) \oplus \ldots \oplus (R/P_k^{n_k}),$$

where the P_i are non-zero maximal ideals of R and the n_i are positive integers. Further, for a given M, the family $\{P_i, n_i\}_{i=1}^k$ is uniquely determined.

To complete the picture, we shall deal in our last result with finitely generated modules which are not torsion modules.

THEOREM 6.20 *Let R be a Dedekind domain and let M be a finitely generated R-module which is not a torsion module. Then*

$$M \approx (R/P_1^{n_1}) \oplus (R/P_2^{n_2}) \oplus \ldots \oplus (R/P_k^{n_k}) \oplus R \oplus R \oplus \ldots \oplus R \oplus I,$$

where:

   (i) *the P_i are non-zero maximal ideals of R;*

   (ii) *the n_i are positive integers;*

   (iii) *there are finitely many summands R;*

   (iv) *I is a non-zero fractional ideal of R.*

Further, the family $\{P_i, n_i\}_{i=1}^k$, the number of summands R and the ideal class of I are determined by M.

Proof. The existence of such an isomorphism follows from Theorem 6.16 and Lemma 6.19. Put

$$M' = (R/P_1^{n_1}) \oplus (R/P_2^{n_2}) \oplus \ldots \oplus (R/P_k^{n_k}) \oplus R \oplus R \oplus \ldots \oplus R \oplus I,$$

so that $M \approx M'$. The torsion submodule of M' consists of all elements which have all but the first k components zero. If we denote by M'' the submodule of M' consisting of all elements which have the first k components zero, then

$$M' = T(M') + M'' \quad \text{(d.s.)}$$

and

$$T(M) \approx T(M') \approx (R/P_1^{n_1}) \oplus (R/P_2^{n_2}) \oplus \ldots \oplus (R/P_k^{n_k}),$$

$$M/T(M) \approx M'/T(M') \approx R \oplus R \oplus \ldots \oplus R \oplus I.$$

It follows from Theorem 6.11 and Lemma 6.18 that the family $\{P_i, n_i\}_{i=1}^k$, the number of summands R and the ideal class of I are uniquely determined by M.□

Exercises on Chapter 6

6.1 A commutative ring which is Noetherian and self-injective is said to be a (commutative) *quasi-Frobenius ring*. Let R be a commutative quasi-local ring. Prove that the following statements are equivalent:

(i) R is quasi-Frobenius;

(ii) R is Artinian and the socle of R is a simple ideal of R.

[*Hint:* Use Exercise 4.15 and Theorem 4.30 to prove that (i) implies (ii); use Proposition 3.17 Corollary, Exercise 5.1 and Theorem 5.21 Corollary to prove that (ii) implies (i).]

6.2 Show that a commutative quasi-Frobenius ring is the direct sum of a finite number of local Artinian quasi-Frobenius rings.

6.3 Show that every non-zero prime ideal of a Dedekind domain is maximal.

6.4 Let R be a non-trivial commutative ring. Show that the following statements are equivalent:

(i) R is a valuation ring (see Exercise 2.7);

(ii) the ideals of R are totally ordered;

(iii) R is quasi-local and the finitely generated ideals of R are principal.

6.5 Let R be a commutative quasi-local domain. Show that R is a Dedekind domain if and only if R is a Noetherian valuation ring.

Notes on Chapter 6

As a generalization of Theorem 6.7, we may ask the following question: which rings R (not necessarily commutative) have the property that every R-module is a direct sum of cyclic modules? This question was raised by G. Köthe. He proved that, if R is a left Artinian principal ideal ring, then it has this property (see [15]). The converse for the case of commutative rings was proved by I. S. Cohen and I. Kaplansky. As a generalization of this, if there is a *set* of modules such that every R-module (R not necessarily commutative) is isomorphic to a direct sum of members of this set, then R is left Artinian. This follows from Theorem 4.4 and a theorem of S. U. Chase [4]. There are no left Artinian rings which are known not to possess such a set of modules.

The structure of finitely generated modules over Dedekind domains was determined by I. Kaplansky.

Referring to Theorem 6.16 Corollary, we may ask which commutative rings R have the property that every finitely generated R-module is a direct sum of cyclic modules. As we have seen, if R is a Noetherian domain, then it must be a principal ideal domain. If we allow non-domains, a similar result is true; if R is a Noetherian ring, then it must be a principal ideal ring (see A. I. Uzkov [23]). But R does not have to be Noetherian; I. Kaplansky has observed that almost maximal valuation domains have this property. E. Matlis has proved a result in the converse direction, namely that a quasi-local domain with this property must be an almost maximal valuation domain. The proofs of these results can be found in D. T. Gill [11], where the results are further extended to non-domains. E. Matlis [18] has also given sufficient conditions for a commutative domain R which is not necessarily quasi-local to have the property that every finitely generated R-module is a direct sum of cyclic modules.

Bibliography

[1] Banaschewski, B. and Bruns, G., Categorical characterization of the MacNeille Completion, *Archiv der Mathem.* **18** (1967), 369–77.

[2] Bass, H., Finitistic homological dimension and a homological generalization of semi-primary rings, *Trans. Amer. Math. Soc.* **95** (1960), 466–88.

[3] Cartan, H. and Eilenberg, S., *Homological Algebra* (Princeton University Press, 1956).

[4] Chase, S. U., Direct products of modules, *Trans. Amer. Math. Soc.* **97** (1960), 457–73.

[5] Faith, C., *Lectures on Injective Modules and Quotient Rings* (Lecture Notes in Mathematics 49, Springer-Verlag, 1967).

[6] Faith, C. and Walker, E., Direct sum representations of injective modules, *J. Algebra* **5** (1967), 203–21.

[7] Freyd, P., *Abelian Categories* (Harper and Row, 1964).

[8] Fuchs, L., *Infinite Abelian Groups*, vol. I (Academic Press, 1970).

[9] Gabriel, P., Des Catégories Abéliennes, *Bull. Soc. Math. France* **90** (1962), 323–448.

[10] Gabriel, P., Objets injectifs dans les catégories abéliennes, *Séminaire Dubreil* **12** (1959), no. 17.

[11] Gill, D. T., Almost maximal valuation rings, *Jour. London Math. Soc.* (to appear).

[12] Halmos, P. R., *Lectures on Boolean Algebras* (Van Nostrand, 1963).

[13] Jans, J. P., *Rings and Homology* (Holt, Rinehart and Winston, 1964).

[14] Kelley, J. L., *General Topology* (Van Nostrand, 1955).

[15] Köthe, G., Verallgemeinerte Abelsche Gruppen mit hyperkomplexen Operatorring, *Math. Z.* **39** (1935), 31–44.

[16] Lambek, J., *Lectures on Rings and Modules* (Ginn Blaisdell, 1966).

[17] Lesieur, L. and Croisot, R., *Algèbre Noethérienne non commutative* (Gauthier–Villars, 1963).

[18] Matlis, E., Decomposable modules, *Trans. Amer. Math. Soc.* **125** (1966), 147–79.

[19] Matlis, E., Injective modules over Noetherian rings, *Pacific J. Math.* **8** (1958), 511–28.

[20] Matlis, E., Injective modules over Prufer rings, *Nagoya Math. Jour.* **15** (1959), 57–69.

[21] Mitchell, B., *Theory of Categories* (Academic Press, 1965).

[22] Northcott, D. G., *Lessons on Rings, Modules and Multiplicities* (Cambridge, 1968).

[23] Uzkov, A. I., On the decomposition of modules over a commutative ring into direct sums of cyclic submodules (in Russian), *Math. Sb.* **62** (1963), 469–75.

[24] Vámos, P., The dual of the notion of 'finitely generated', *J. London Math. Soc.* **43** (1968), 643–6.

Index